SPRINGER PROTOCOLS HANDBOOKS

For further volumes:
http://www.springer.com/series/8623

Springer Protocols Handbooks collects a diverse range of step-by-step laboratory methods and protocols from across the life and biomedical sciences. Each protocol is provided in the Springer Protocol format: readily-reproducible in a step-by-step fashion. Each protocol opens with an introductory overview, a list of the materials and reagents needed to complete the experiment, and is followed by a detailed procedure supported by a helpful notes section offering tips and tricks of the trade as well as troubleshooting advice. With a focus on large comprehensive protocol collections and an international authorship, Springer Protocols Handbooks are a valuable addition to the laboratory.

Analytical Techniques in Biochemistry

Mahin Basha

P. M. Sayeed Calicut University Centre, University of Calicut, Lakshadweep, India

Mahin Basha
P. M. Sayeed Calicut University Centre
University of Calicut
Lakshadweep, India

ISSN 1949-2448   ISSN 1949-2456   (electronic)
Springer Protocols Handbooks
ISBN 978-1-0716-0133-4   ISBN 978-1-0716-0134-1   (eBook)
https://doi.org/10.1007/978-1-0716-0134-1

This Humana imprint is published by the registered company Springer Science+Business Media, LLC, part of Springer Nature.
The registered company address is: 233 Spring Street, New York, NY 10013, U.S.A.

Contents

About the Author

MAHIN BASHA, MSc (Biochemistry), MTech (Industrial Biotechnology), and PhD (Chemical Engineering), has over 7 years of experience in the field of academic research and teaching at various respected institutions. He is currently serving as Assistant Professor of Biochemistry at P. M. Sayeed Calicut University Centre, affiliated to University of Calicut, Kerala, India, one of the India's most prestigious institutes. Dr. Basha's research has been published in a number of leading international journals, and he has presented his work at scientific meetings around the globe.

Chapter 1

Sedimentation

Abstract

Sedimentation is the process of letting suspended material settle by gravity. It is accomplished by decreasing the velocity of the water being treated to a point in which the particles will no longer remain in suspension. When the velocity no longer supports the transport of the particles, gravity will remove them from the flow.

For example, in a glass cylinder, when solids diffuse through the interface, the process starts then to settle from slurry during a batch settling test and is assumed to approach terminal velocities under hindered settling conditions. Thus, several zones of concentration will be established. The particle is not actually sent all the way to the bottom of the cell, resulting in sediment. Rather, a low centrifugal field is used to create a concentration gradient wherein more particles are near the bottom of the cell than near the top. When the temperature decreases, the rate of settling becomes slower.

Key words Sedimentation, Gravity sedimentation, Zone settling velocity, Stokes' law

Abbreviation

ZSV   Zone settling velocity

Definition:
Sedimentation is the process of allowing particles in suspension in water to settle out of the suspension under the effect of gravity to the bottom. The particles that settle out from the suspension become sediment; in water treatment this sediment is known as sludge.

Sedimentation is affected by particle concentration. From everyday experience, the effect of sedimentation due to the influence of the Earth's gravitational field (g ¼ 981 cm s^{-2}) versus the increased rate of sedimentation in a centrifugal field (g > 981 cm s^{-2}) is apparent. To give a simple but illustrative example, crude sand particles added to a bucket of water travel slowly to the bottom of the bucket by gravitation, but they sediment much faster when the bucket is swung around in a circle. Similarly, biological structures exhibit a drastic increase in sedimentation when they undergo acceleration in a centrifugal field. The relative centrifugal field is usually expressed as a multiple of the acceleration due to gravity [1].

Mahin Basha, *Analytical Techniques in Biochemistry*, Springer Protocols Handbooks, https://doi.org/10.1007/978-1-0716-0134-1_1, © Springer Science+Business Media, LLC, part of Springer Nature 2020

- In dilute suspensions, particles act independently.
- In concentrated suspensions, particle–particle interactions are significant.
  - Particles may collide and stick together (form flocs).
  - Particle flocs may settle more quickly.
  - At very high concentrations, particle–particle forces may prevent further consolidation.

Most wastewaters and waters contain solids, and in many treatment processes, solids are generated, e.g., phosphate precipitation, coagulation, and activated sludge bio-oxidation. Particles in water and wastewater that will settle by gravity within a reasonable period of time can be removed by "sedimentation" in sedimentation basins (also known as "clarifiers").

"Settleable" does not necessarily mean that these particles will settle easily by gravity. In many cases they must be coaxed out of suspension or "solution" by the addition of chemicals or increased gravity (centrifugation or filtration).

Because of the high volumetric flow rates associated with water and wastewater treatment systems, gravity sedimentation is the only practical, economical method to remove these solids, i.e., processes such as centrifugation are not economical, in most cases.

Gravity separation can obviously be applied only to those particles which have density greater than water. But this density must be significantly greater than that of water due to particle surface effects and turbulence in the sedimentation tanks.

1   Gravity Sedimentation

Gravitational sedimentation methods of particle size determination are based on the settling behavior of a single sphere, under gravity, in a fluid of infinite extent. Unique relationships between settling velocity and particle size and between drag factor and Reynolds number have been found:

- Produce a clarified (free of suspended solids) effluent
- Produce a highly concentrated solid sludge stream

Types of Settling: *Four types of sedimentation:*

1. *Discrete settling*
2. *Flocculent settling*
3. *Hindered settling*
4. *Compression*

1.1 Type I (Discrete Sedimentation)

Unhindered settling is a process that removes the discrete particles in a very low concentration without interference from nearby particles. In general, if the concentration of the solutions is lower than 500 mg/L total suspended solids, sedimentation will be considered discrete.

- Occurs in dilute suspensions whose particles have very little interaction with each other as they settle.
- Particles settle according to Stokes' law.
- Design parameter is surface overflow rate.
    Example: Removal of grit and sand in wastewater treatment

1.2 Type II (Flocculent Sedimentation)

It is a process wherein colloids come out of suspension in the form of floc, either spontaneously or due to the addition of a clarifying agent. It is used in applications like water purification, sewage treatment, cheese production, and brewing (Fig. 1).

- Particles flocculate as they settle.
- Floc particle velocity increases with time.

Design parameters:

1. Surface overflow rate
2. Depth of tank
3. Hydraulic retention time

1.3 Zone Settling (Type III)

Zone settling occurs when a flocculent suspension with high initial concentration (on the order of 500 mg/L) settles by gravity. Flocculent forces between particles cause settling as a matrix (particles remain in a fixed position relative to each other as they settle).

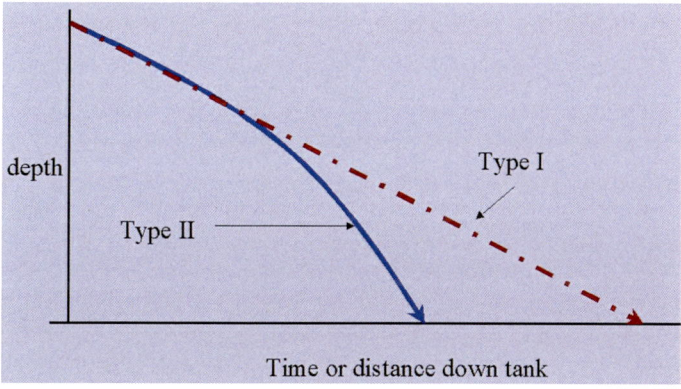

Fig. 1 Comparison of Type I and II sedimentation

1.4 Compression (Type IV)

When matrix sedimentation is constrained from the bottom, the matrix begins to compress. Such a situation occurs when the matrix encounters the bottom of tank in which it is settling. This is called compression (Type IV) settling as shown in Fig. 2.

The height of the interface (between the clarified zone and the zone settling zone) versus time is plotted in the figure below to determine the "zone settling velocity" (ZSV). Velocity of this interface is steady after some induction period but changes with time as compression begins. The slope of the steady interface subsidence rate represents zone settling velocity as shown in Fig. 3.

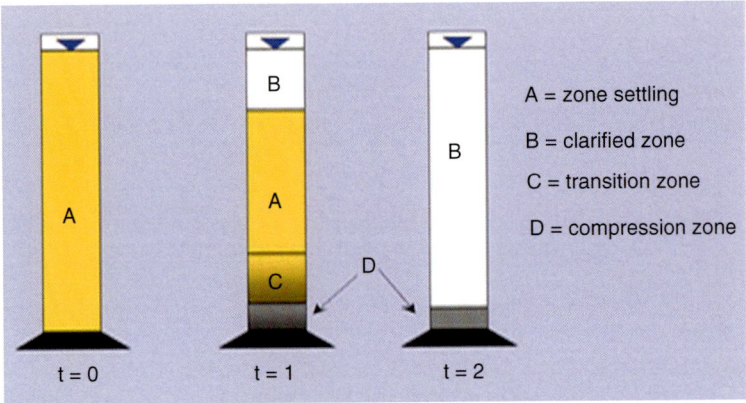

Fig. 2 Settling types are demonstrated in a batch settling test as illustrated

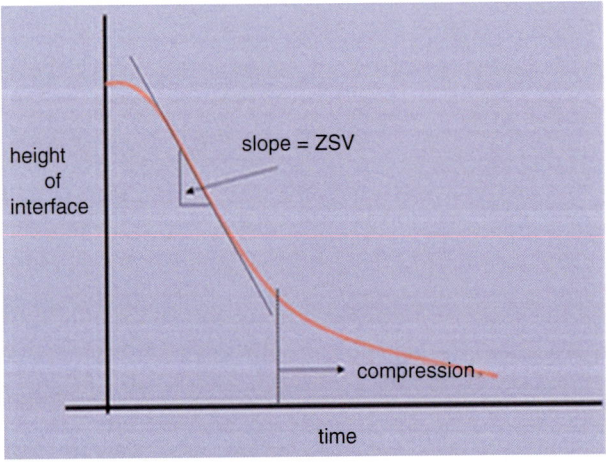

Fig. 3 The slope of the steady interface subsidence rate represents zone settling velocity

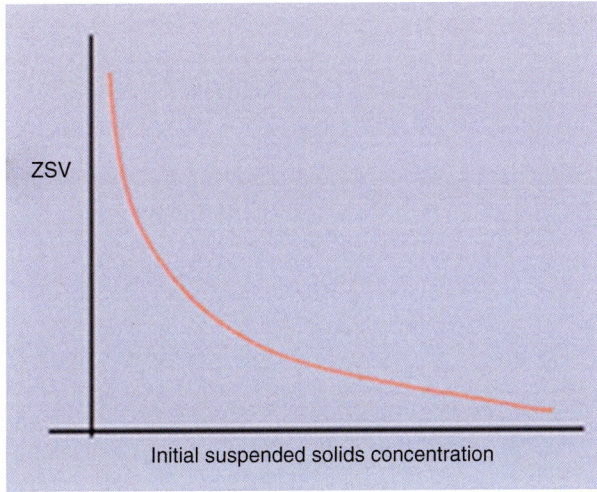

Fig. 4 Relationship between initial suspended solids and ZSV

Initial suspended solids concentration has a significant effect on the ZSV because the higher the suspended solids concentration, the more difficult it is to pass water through the pore spaces in the settling matrix (the only way a matrix can settle is if the water below it is allowed to pass upward through the matrix). A typical relationship between initial suspended solids and ZSV is shown in Fig. 4.

Factors affecting zone settling velocity:

1. Suspended solids concentration
2. Depth of settling column (or tank)
3. Stirring (0.5–2 rpm to prevent "arching")
4. Temperature
5. Polymer addition (affects matrix structure)

Two important functions of these sedimentation tanks are *clarification* and *thickening.* For a continuous flow clarifier, operated at steady state, mass flow of suspended solids can schematically represent as shown in Fig. 5:

X = Influent suspended solids concentration

X_e = Effluent suspended solids concentration (often close to zero)

X_u = Underflow (thickened) suspended solids concentration

Q = Influent volumetric flow rate

Q_u = Underflow volumetric flow rate

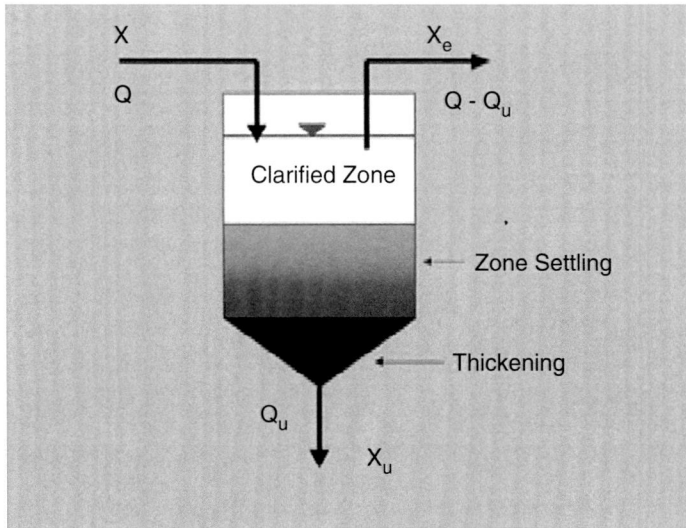

Fig. 5 Schematically represent clarification and thickening

Batch Flux Method: The batch flux method is one way to analyze and select design parameters for the clarifiers/thickeners. Start by considering the mass flux of solids through the clarifier/thickener. There are two components of this flux:

1. Subsidence (sedimentation)
2. Bulk transport (due to sludge withdrawal from bottom of the tank)
   Total flux of solids through the clarifier is given by:

$$G = v_i X_i + u X_i$$

Where:

G = Mass flux (mass of SS transported/area-time)

V_i = Zone settling velocity (ZSV) at X_i

u = Bulk transport velocity due to sludge withdrawal from bottom of the tank

$$u = Q_u/A_s$$

Q_u = Underflow rate (withdrawal rate)

A_s = Cross-sectional area of clarifier

Zone settling velocity is highly dependent on X_i, so to calculate the flux due to subsidence, we need to assume a typical relationship (as shown above) between zone settling velocity and X_i to get:

Solid flux due to subsidence (settling) is calculated by:

$$G_s = (v_i)(x_i)(\text{mass/time-area})$$

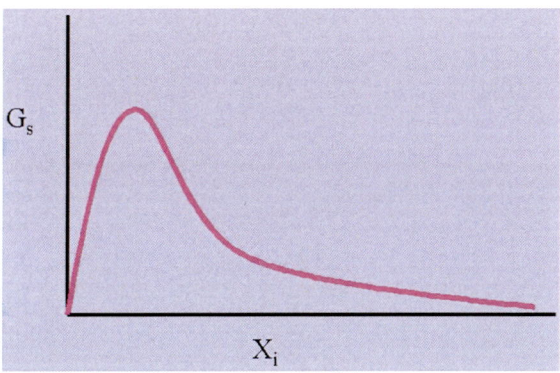

For a particular underflow rate u, there is a minimum in the flux capacity of the clarifier. This minimum occurs at $X_i = X_L$ (note there is also a minimum G at the origin, but this has no relevance since even after the influent X is diluted X_i never gets this low). Therefore, for a given underflow rate, there is a "limiting flux" which can be transmitted through the clarifier. As X_i passes from X_f (suspended solids concentration in the influent) to X_u, it must pass through this bottleneck $X_i = X_L$. This controls the solids loading rate to the clarifier. Essentially, for a critically loaded clarifier, there exist only two suspended solids concentrations, X_L and X_A, if the compression zone is ignored. An explanation of "two-concentration" critically loaded clarifier follows. Suspended solids enter the clarifier at some initial concentration X_f. These solids are diluted by clarified effluent. As the solids settle, they concentrate and ultimately reach X_L. Suspended solids cannot be transmitted as fast through this layer as in the layers above (because the influent has lower suspended solids concentration and therefore higher zone settling velocity) so there is a buildup of suspended solids at X_L. At steady state the influent suspended solids have to be diluted to X_A to balance fluxes through the clarifier (at steady state all the solids fluxes must be equal at all depths). Any other concentration will cause the layers to disappear, either by washing out over the effluent or by being drawn through the bottom of the clarifier.

When the clarifier is critically loaded, i.e., when the loading rate equals the flux capacity of the clarifier, the resultant concentration profile in the clarifier is given by:

The batch settling data can be represented by an exponential function. For example, the following equation is an exponential curve fit to the settling data same as shown in Fig. 6:

$$V = 200\,\frac{\text{m}}{\text{h}} \cdot e^{-0.4\frac{\text{liter}}{\text{g}}\cdot X}$$

$$X = \text{g/liter}$$

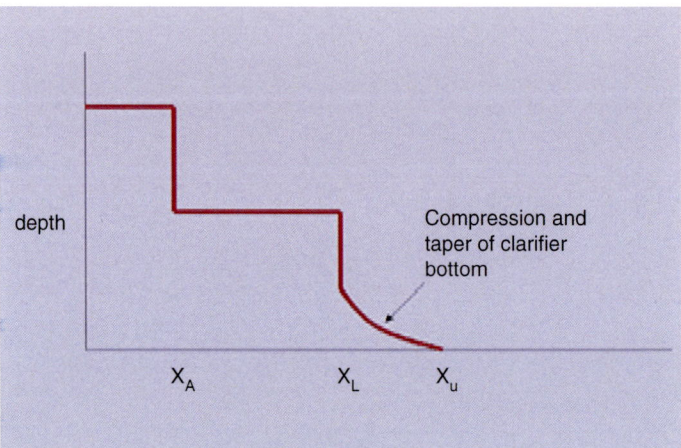

Fig. 6 Exponential curve fit to the settling data

Applications: Stokes' law can be used to determine the surface area of a settling tank:

– Set the critical velocity equal to the settling velocity of the smallest particle.

– The overflow rate is equal to the flow rate into the tank divided by the surface area.

– Setting the overflow rate equal to the critical settling velocity allows time to capture smallest particles of interest.

$$\text{OFR} = v_c = \frac{Q}{A}$$

OFR = Overflow rate (m/s or ft/s)

v_c = Critical settling velocity (m/s or ft/s)

Q = The flow rate into the basin (m^3/s or ft^3/s)

A = The surface area of the basin (m^2 or ft^2)

Reference

1. Rajan K (2011) Analytical techniques in biochemistry and molecular biology. Springer, New York. eBook ISBN 978-1-4419-9785-2

Chapter 2

Filtration

Abstract

The principle of filtration operation is to remove the suspended particulates in water by passing the water through a membrane of porous material. Filtration is used to remove suspended matter in general such as clays, algae, and asbestos fibers from water. It is possible to carry out filtration under a variety of conditions, but a number of factors influence filtration. In filtration a filter cake gradually builds up as the filtrate passes through the filter cloth. As the filter cake increases in thickness, the resistance to flow gradually increases. Alternatively, if the flow rate is to be kept constant, the pressure will gradually have to be increased. The flow rate may also be reduced by blocking of holes in the filter cloth. This chapter deals with the outline of micro and ultrafiltration.

Key words Filtration, Microfiltration, Ultrafiltration

Filtration is any of various mechanical, physical, or biological operations that separate solids from fluids (liquids or gases) by adding a filter medium through which only the fluid can pass. The fluid that passes through is called the filtrate [1].

1 Types of Filtration

Type1: Cross-Flow Filtration:
Flow parallel to membrane surface does not cause buildup and therefore does not suffer from reduced flow over time as shown in Fig. 1.

F = Feed

M = Membrane

P = Permeate

R = Retention (components that do NOT pass through the membrane)

Mahin Basha, *Analytical Techniques in Biochemistry*, Springer Protocols Handbooks,
https://doi.org/10.1007/978-1-0716-0134-1_2, © Springer Science+Business Media, LLC, part of Springer Nature 2020

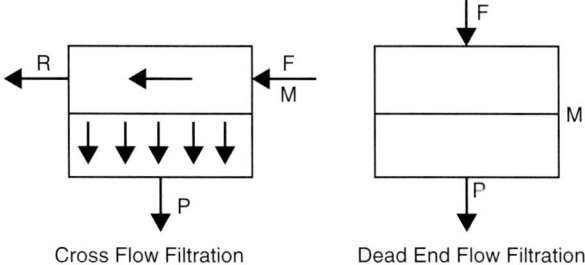

Cross Flow Filtration   Dead End Flow Filtration

Fig. 1 Cross-flow and dead-end flow filtration

Type 2: Dead-End Flow Filtration:
Flow perpendicular to membrane surface causes buildup of filter cake on membrane as shown in Fig. 1.

F = Feed

M = Membrane

P = Permeate (components that pass through membrane)

2 Microfiltration

It separates soluble contaminants remaining within the supernatant. Supernatant may contain proteins, biomolecules, and unused growth media.

Microfiltration works based on a pressure-driven process. It separates components in a solution or suspension based on molecular size. Particles sizes range from 10 mm (starches) to approx. 0.04 mm (DNA, viruses, and globular proteins) as shown in Fig. 2.

3 Ultrafiltration

It is usually used to further separate any contaminants able to pass through the microfiltration membrane using a pressure gradient as shown in Fig. 3.

It separates particles whose sizes range from 0.1 mm to 0.001 mm.

Usually based on molecular weight, typical range is 200–300,000 g/mole (Table 1).

Applications:

1. Filtration is used to separate particles and fluid in a suspension, where the fluid can be a liquid, a gas, or a supercritical fluid. Depending on the application, either one or both of the components may be isolated.

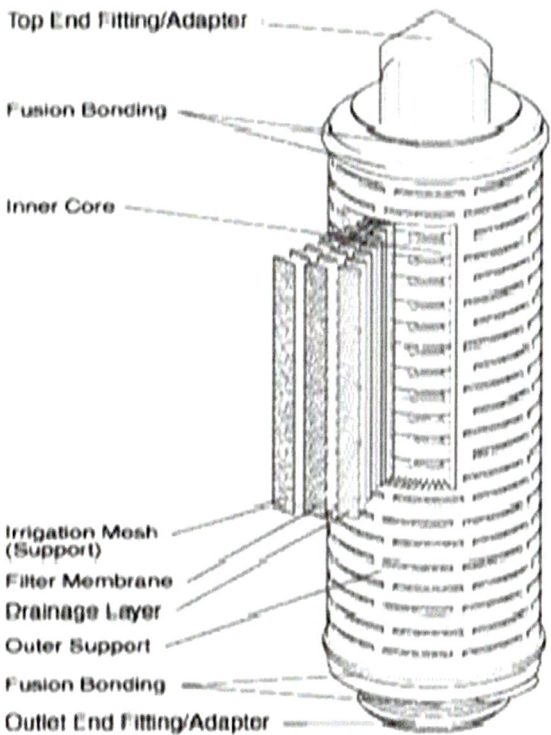

Top End Fitting/Adapter

Fusion Bonding

Inner Core

Irrigation Mesh
(Support)

Filter Membrane

Drainage Layer

Outer Support

Fusion Bonding

Outlet End Fitting/Adapter

Fig. 2 Microfiltration image

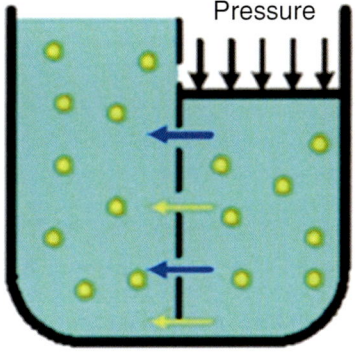

Pressure

(Solution moves by
pressure gradient)

Fig. 3 Ultrafiltration

Table 1
Microfiltration vs. ultrafiltration

S. no.	Microfiltration	Ultrafiltration
1.	Proteins act as the permeate	Proteins act as the retentate
2.	Separates larger particles, for example, colloids, fat globules, and cells	Separates smaller particles, for example, macromolecules
3.	Particle size range: 10mm (starches) to approximately 0.04 mm (DNA, viruses, and globular proteins)	Particle size range: 0.1 mm to 0.001 mm

2. Filtration, as a physical operation, is very important in chemistry for the separation of materials of different chemical compositions.

3. Filtration is also important and widely used as one of the unit operations of chemical engineering.

4. Filtration differs from sieving, where separation occurs at a single perforated layer (a sieve). In sieving, particles that are too big to pass through the holes of the sieve are retained (see particle size distribution). In filtration, a multilayer lattice retains those particles that are unable to follow the tortuous channels of the filter. Oversized particles may form a cake layer on top of the filter and may also block the filter lattice, preventing the fluid phase from crossing the filter (blinding).

5. Filtration differs from adsorption, where it is not the physical size of particles that causes separation but the effects of surface charge. Some adsorption devices containing activated charcoal and ion exchange resin are commercially called filters, although filtration is not their principal function.

Reference

1. Rajan K (2011) Analytical techniques in biochemistry and molecular biology. Springer, New York. eBook ISBN978-1-4419-9785-2

Chapter 3

Centrifugation

Abstract

Biological centrifugation is a process that uses centrifugal force to separate and purify mixtures of biological particles in a liquid medium. It is a key technique for isolating and analyzing cells, subcellular fractions, supramolecular complexes, and isolated macromolecules such as proteins or nucleic acids. The development of the first analytical ultracentrifuge by Svedberg in the late 1920s and the technical refinement of the preparative centrifugation technique by Claude and colleagues in the 1940s positioned centrifugation technology at the center of biological and biomedical research for many decades. Today, centrifugation techniques represent a critical tool for modern biochemistry and are employed in almost all invasive subcellular studies. While analytical centrifugation is mainly concerned with the study of purified macromolecules or isolated supramolecular assemblies, preparative centrifugation methodology is devoted to the actual separation of tissues, cells, subcellular structures, membrane vesicles, and other particles of biochemical interest.

Key words Centrifugation, Differential centrifugation, Density-gradient centrifugation, Ultracentrifuge

Abbreviation

RPM   Revolutions per minute

Most undergraduate students will be exposed to preparative centrifugation protocols during practical classes and might also experience a demonstration of analytical centrifugation techniques. This chapter gives theoretical background of centrifugation, an overview of practical aspects of using centrifuges in the biochemical laboratory, an outline of preparative centrifugation, and a description of the usefulness of ultracentrifugation techniques in the biochemical characterization of macromolecules [1, 2].

Separating and purifying biological materials can be performed by filtration and centrifugation. Chromatography and electrophoresis are also common methods.

Principle: *The rate of settling of a particle, or the rate of separation of two immiscible liquids, is increased many times by the application of a centrifugal field (force) many times that of gravity as shown in Fig. 1.*

Mahin Basha, *Analytical Techniques in Biochemistry*, Springer Protocols Handbooks,
https://doi.org/10.1007/978-1-0716-0134-1_3, © Springer Science+Business Media, LLC, part of Springer Nature 2020

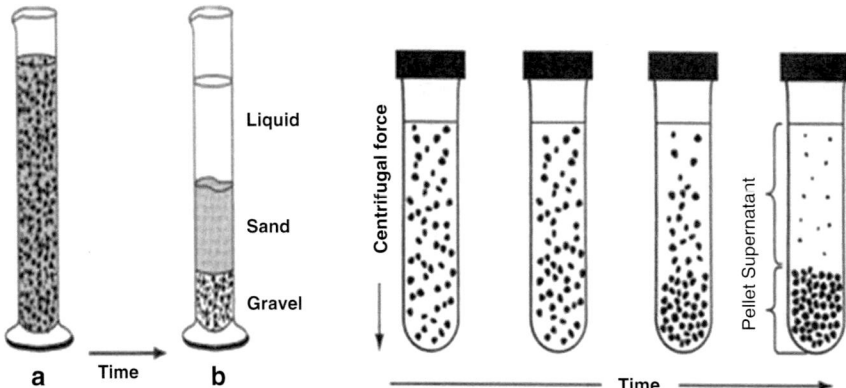

Fig. 1 Filtration by centrifugation

Applications:

- Separate two immiscible liquids
- Isolate cellular organelles
- Isolate DNA, RNA, and proteins
- Isolate small particles including:
  - Bacteria
  - Viruses
  - Cells

Force in a centrifuge is proportional to two things: first, it depends on how fast the centrifuge spins and, second, it depends on the radius of rotation – think about "crack the whip" as shown in Fig. 2.

Relative Centrifugal Force (RCF):

Also = Xg

$$RCF = 11.17(r)(n/1000)^2$$

Where r = radius in cm from centerline and n = rotor speed in RPM or revolutions per minute

Particle sedimentation depends on RCF in the centrifuge, size of the particle, particle density, liquid density, and liquid viscosity.

If a particle has the same density as the liquid around it, the particle does not move. If a particle is denser than the liquid, it moves down the tube. If a particle is less dense than the liquid, it moves up!

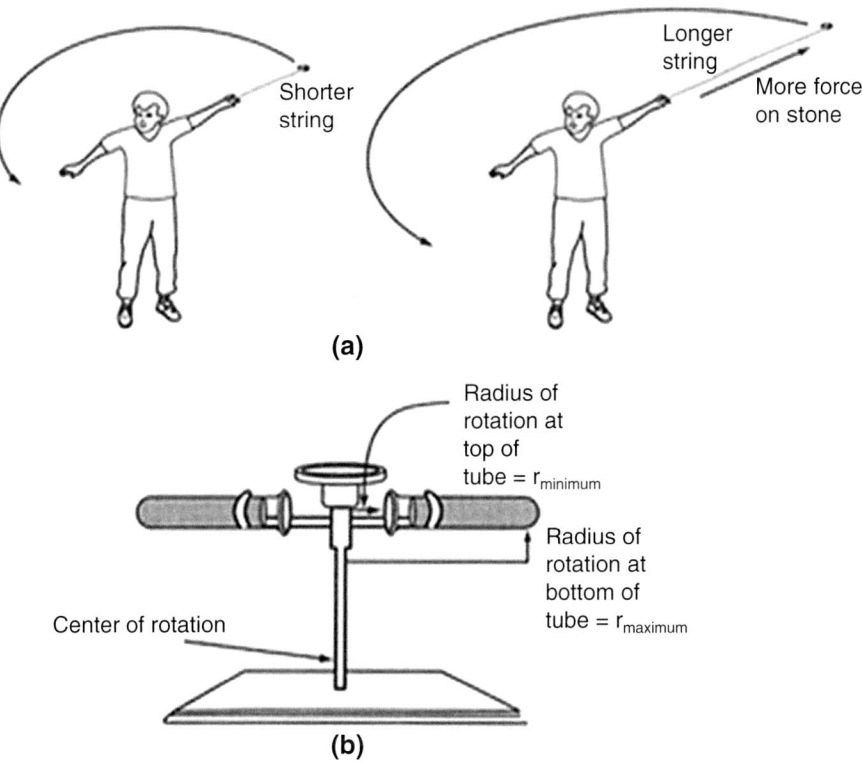

(a)

(b)

Fig. 2 The radius of rotation. (**a**) The longer a string being whirled, the greater the radius of rotation and the more force that is experienced by the store. (**b**) In a centrifuge, the further a particle is from the center of rotation, the more force it experiences

1 Two Basic Modes of Centrifugation

Centrifugation techniques take a central position in modern biochemical, cellular, and molecular biological studies. Depending on the particular application, centrifuges differ in their overall design and size. However, a common feature in all centrifuges is the central motor that spins a rotor containing the samples to be separated. Particles of biochemical interest are usually suspended in a liquid buffer system contained in specific tubes or separation chambers that are located in specialized rotors. The biological medium is chosen for the specific centrifugal application and may differ considerably between preparative and analytical approaches. As outlined below, the optimum pH value, salt concentration, stabilizing cofactors, and protective ingredients such as protease inhibitors have to be carefully evaluated in order to preserve biological function [1, 2].

Some large-volume centrifuge models are quite demanding on space and also generate considerable amounts of heat and noise and are therefore often centrally positioned in special instrument rooms in biochemistry departments. However, the development of small-capacity bench-top centrifuges for biochemical applications, even in the case of ultracentrifuges, has led to the introduction of these models in many individual research laboratories.

The main wtypes of centrifuge encountered by undergraduate students during introductory practicals may be divided into microfuges (so called because they centrifuge small-volume samples in Eppendorf tubes), large-capacity preparative centrifuges, high-speed refrigerated centrifuges, and ultracentrifuges. Simple bench-top centrifuges vary in design and are mainly used to collect small amounts of biological material, such as blood cells. To prevent denaturation of sensitive protein samples, refrigerated centrifuges should be employed. Modern refrigerated microfuges are equipped with adapters to accommodate standardised plastic tubes for the sedimentation of $0.5–1.5$ cm^3 volumes. They can provide centrifugal fields of approximately 10,000 g and sediment biological samples in minutes, making microfuges an indispensable separation tool for many biochemical methods. Microfuges can also be used to concentrate protein samples. For example, the dilution of protein samples, eluted by column chromatography, can often represent a challenge for subsequent analyses. Accelerated ultrafiltration with the help of plastic tube-associated filter units, spun at low g-forces in a microfuge, can overcome this problem. Depending on the proteins of interest, the biological buffers used, and the molecular mass cut-off point of the particular filters, a 10- to 20-fold concentration of samples can be achieved within minutes. Larger preparative bench-top centrifuges develop maximum centrifugal fields of 3000–7000 g and can be used for the spinning of various types of containers. Depending on the range of available adapters, considerable quantities of $5–250$ cm^3 plastic tubes or 96-well ELISA plates can be accommodated. This gives simple and relatively inexpensive bench-top centrifuges a central place in many high-throughput biochemical assays where the quick and efficient separation of coarse precipitates or whole cells is of importance [1, 2].

High-speed refrigerated centrifuges are absolutely essential for the sedimentation of protein precipitates, large intact organelles, and cellular debris derived from tissue homogenization and microorganisms.

They operate at maximum centrifugal fields of approximately 100,000 g. Such centrifugal force is not sufficient to sediment smaller microsomal vesicles or ribosomes, but can be employed to differentially separate nuclei, mitochondria, or chloroplasts. In addition, bulky protein aggregates can be sedimented using high-speed refrigerated centrifuges. An example is the contractile apparatus released from muscle fibers by homogenization, mostly consisting of myosin and actin macromolecules aggregated in filaments. In order to harvest yeast cells or bacteria from large volumes of culture media, high-speed centrifugation may also be used in a continuous flow mode with zonal rotors. This approach does not therefore use centrifuge tubes but a continuous flow of medium. As the medium enters the moving rotor, biological particles are sedimented against the rotor periphery and excess liquid is removed through a special outlet port. Ultracentrifugation has decisively advanced the detailed biochemical analysis of subcellular structures and isolated biomolecules. Preparative ultracentrifugation can be operated at relative centrifugal fields of up to 900,000 g. In order to minimize excessive rotor temperatures generated by frictional resistance between the spinning rotor and air, the rotor chamber is sealed, evacuated, and refrigerated. Depending on the type, age, and condition of a particular ultracentrifuge, cooling to the required running temperature and the generation of a stable vacuum might take a considerable amount of time. To avoid delays during biochemical procedures involving ultracentrifugation, the cooling and evacuation system of older centrifuge models should be switched on at least an hour prior to the centrifugation run. On the other hand, modern ultracentrifuges can be started even without a fully established vacuum and will proceed in the evacuation of the rotor chamber during the initial acceleration process. For safety reasons, heavy armor plating encapsulates the ultracentrifuge to prevent injury to the user in case of uncontrolled rotor movements or dangerous vibrations. A centrifugation run cannot be initiated without proper closing of the chamber system. To prevent unfavorable fluctuations in chamber temperature, excessive vibrations, or operation of rotors above their maximum rated speed, newer models of ultracentrifuges contain sophisticated temperature regulation systems, flexible drive shafts, and an over-speed control device. Although slight rotor imbalances can be absorbed by modern ultracentrifuges, a more severe misbalance of tubes will cause the centrifuge to switch off automatically. This is especially true for swinging-bucket rotors [1, 2].

The most familiar methods are differential centrifugation and density-gradient centrifugation as shown in Fig. 3a, b.

Desktop or clinical centrifuges, <10,000 RPM

High-speed centrifuges, 10,000–30,000 RPM (around 50,000 Xg)

Ultracentrifuges, up to 80,000 RPM (500,000 Xg)

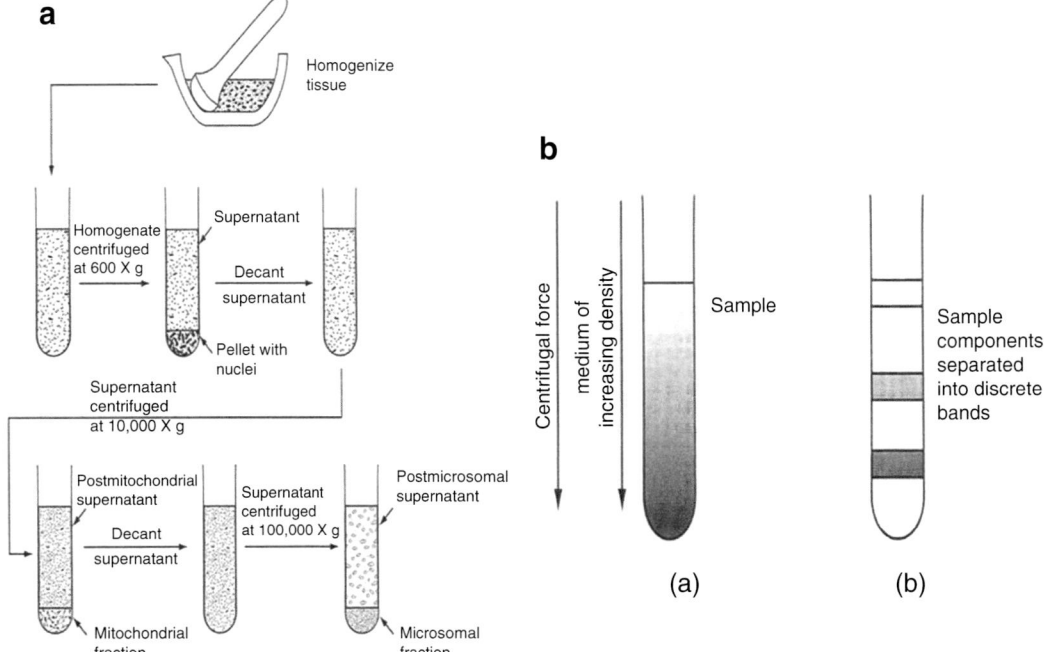

Fig. 3 (**a**) Differential centrifuge, (**b**) Density-gradient centrifuge

2   Instrument Design

Safety: *Centrifuges look sturdy, sort of like washing machines. But, they are probably the most dangerous instrument any of you will use; also they are surprisingly easy to damage. Rotors must withstand huge forces. In an ultracentrifuge, a 1 g particle "weighs" 0.65 tons; any imperfection will weaken the rotor. Modern centrifuges are not only highly sophisticated but also relatively sturdy pieces of biochemical equipment that incorporate many safety features. Rotor chambers of high-speed and ultracentrifuges are always enclosed in heavy armor plating. Most centrifuges are designed to buffer a certain degree of imbalance and are usually equipped with an automatic switch-off mode. However, even in a well-balanced rotor, tube cracking during a centrifugation run might cause severe imbalance resulting in dangerous vibrations. When the rotor can only be partially loaded, the order of tubes must be organized according to the manufacturer's instructions, so that the load is correctly distributed. This is important not only for ultracentrifuges with enormous centrifugal fields but also for both small- and large-capacity bench-top centrifuges where the rotors are usually mounted on a more rigid suspension. When using swinging-bucket rotors, it is important always to load all buckets with their caps properly screwed on. Even if only two tubes are loaded with* *solutions, the empty swinging buckets also have to be assembled since they form an integral part of the overall balance of the rotor system. In some swinging-bucket rotors, individual rotor buckets are numbered and should not be interchanged between their designated positions on similarly numbered hinge pins. Centrifugation runs using swinging-bucket rotors are usually set up with low acceleration and deceleration rates, as to avoid any disturbance of delicate gradients and reduce the risk of disturbing bucket attachment. This practice also avoids the occurrence of sudden imbalances due to tube deformation or cracking and thus eliminates potentially dangerous vibrations. If centrifugal separation processes have to be performed routinely with a potentially harmful substance, it makes sense to dedicate a particular centrifuge and accompanying rotors for this work and thereby eliminate the potential of cross-contamination.*

Care and Maintenance of Centrifuges:
Corrosion and degradation due to biological buffer systems used within rotors or contamination of the interior or exterior of the centrifuge via spillage may seriously affect the lifetime of this equipment. Another important point is the proper balancing of centrifuge tubes. This is not only important with respect to safety, as outlined below, but might also cause vibration-induced damage to the rotor itself and the drive shaft of the centrifuge. Thus, proper handling and care, as well as regular maintenance, of both centrifuges and rotors is an important part of keeping this biochemical method available in the laboratory. In order to avoid damaging the protective layers of rotors, such as polyurethane paint or aluminum oxide, care should be taken in the cleaning of the rotor exterior. Coarse brushes that may scratch the finish should not be used and only non-corrosive detergents should be employed. Corrosion may be triggered by long-term exposure of rotors to alkaline solutions, acidic buffers, aggressive detergents, or salt. Thus, rotors should be thoroughly washed with distilled or deionized water after every run. For overnight storage, rotors should be first left upside down to drain excess liquid and then positioned in a safe and dry place. To avoid damage to the hinge pins of swinging-bucket rotors, they should be dried with tissue paper following removal of biological buffers and washing with water. Centrifuge rotors are often not properly stored in a clean environment; this can quickly lead to the destruction of the protective rotor coating and should thus be avoided. It is advisable to keep rotors in a special clean room.

Differential Centrifugation:
Cellular and subcellular fractionation techniques are indispensable methods used in biochemical research. Although the proper separation of many subcellular structures is absolutely dependent on preparative ultracentrifugation, the isolation of large cellular structures, the nuclear fraction, mitochondria, chloroplasts, or large protein precipitates can be achieved by conventional high-speed

refrigerated centrifugation. Differential centrifugation is based upon the differences in the sedimentation rate of biological particles of different size and density. Crude tissue homogenates containing organelles, membrane vesicles, and other structural fragments are divided into different fractions by the stepwise increase of the applied centrifugal field. Following the initial sedimentation of the largest particles of a homogenate (such as cellular debris) by centrifugation, various biological structures or aggregates are separated into pellet and supernatant fractions, depending upon the speed and time of individual centrifugation steps and the density and relative size of the particles. To increase the yield of membrane structures and protein aggregates released, cellular debris pellets are often rehomogenized several times and then recentrifuged. This is especially important in the case of rigid biological structures such as muscular or connective tissues or in the case of small tissue samples as is the case with human biopsy material or primary cell cultures. Resulting supernatant fractions are centrifuged at a higher speed and for a longer time to separate medium-sized and small-sized particles. With respect to the separation of organelles and membrane vesicles, crude differential centrifugation techniques can be conveniently employed to isolate intact mitochondria and microsomes.

Density-Gradient Centrifugation:
To further separate biological particles of similar size but differing density, ultracentrifugation with preformed or self-establishing density gradients is the method of choice. Either rate separation or equilibrium methods can be used. In Fig. 4, the preparative

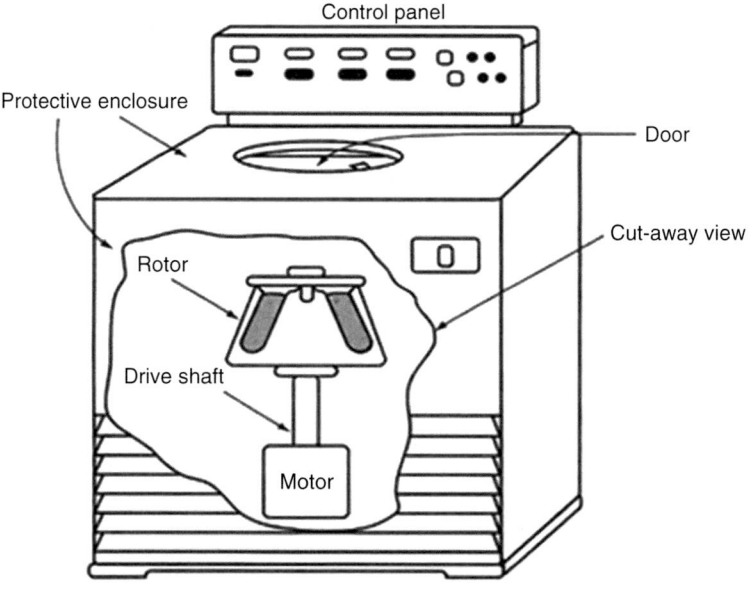

Fig. 4 The figure shows basic components of the centrifuge

ultracentrifugation of low- to high-density particles is shown. A mixture of particles, such as that present in a heterogeneous microsomal membrane preparation, is layered on top of a preformed liquid density gradient. Depending on the particular biological application, a great variety of gradient materials are available. Cesium chloride is widely used for the banding of DNA and the isolation of plasmids, nucleoproteins, and viruses. Sodium bromide and sodium iodide are employed for the fractionation of lipoproteins and the banding of DNA or RNA molecules, respectively. Various companies offer a range of gradient material for the separation of whole cells and subcellular particles, e.g., Percoll, Ficoll, dextran, metrizamide, and Nycodenz. For the separation of membrane vesicles derived from tissue homogenates, ultra-pure DNase-, RNase-, and protease-free sucrose represents a suitable and widely employed medium for the preparation of stable gradients. If one wants to separate all membrane species spanning the whole range of particle densities, the maximum density of the gradient must exceed the density of the most dense vesicle species. Both step gradient and continuous gradient systems are employed to achieve this. If automated gradient makers are not available, which is probably the case in most undergraduate practical classes, the manual pouring of a stepwise gradient with the help of a pipette is not so time-consuming or difficult. In contrast, the formation of a stable continuous gradient is much more challenging and requires a commercially available gradient maker. Following pouring, gradients are usually kept in a cold room for temperature equilibration and are moved extremely slowly in special holders so as to avoid mixing of different gradient layers. For rate separation of subcellular particles, the required fraction does not reach its isopycnic position within the gradient. For isopycnic separation, density centrifugation is continued until the buoyant density of the particle of interest and the density of the gradient are equal.

References

1. Rajan K (2011) Analytical techniques in biochemistry and molecular biology. Springer, New York. eBook ISBN 978-1-4419-9785-2

2. Wilson K, Walker J (2010) Principles and techniques of biochemistry and molecular biology. Cambridge University Press, Cambridge. ISBN 978-0-521-73167-6

Chapter 4

Colorimetry and Spectrophotometer (Spectrophotometry)

Abstract

Spectroscopic techniques employ light to interact with matter and thus probe certain features of a sample to learn about its consistency or structure. Light is electromagnetic radiation, a phenomenon exhibiting different energies, and dependent on that energy, different molecular features can be probed. The basic principles of interaction of electromagnetic radiation with matter are treated in this chapter. There is no obvious logical dividing point to split the applications of electromagnetic radiation into parts treated separately. The justification for the split presented in this text is purely pragmatic and based on "common practice." The applications considered in this chapter use visible or UV light to probe consistency and conformational structure of biological molecules.

Key words Colorimetry, Spectroscopy, UV–Vis spectroscopy, Beer's law

Abbreviations

Ab   Absorbance
SOP   Standard operating procedure

1   Introduction

The light that we see is just a small region of electromagnetic spectrum, to which human eyes are capable of sensing. This region is called "visible light."

This is due to fusion reaction in the sun that occurs when two hydrogen atoms fuse with each other to form a helium molecule. In this process, some mass is removed in the form of energy. This energy is the electromagnetic radiation. This is partly electric and partly magnetic, so it is called as electromagnetic radiation. When the visible light is reflected from any object and touches the sensors of our eyes, we can see the objects. In the process of reflection out of multiple colors of the radiation, some colors are absorbed and some are reflected. The color that is reflected is the color that we can see [1, 2].

Mahin Basha, *Analytical Techniques in Biochemistry*, Springer Protocols Handbooks, https://doi.org/10.1007/978-1-0716-0134-1_4, © Springer Science+Business Media, LLC, part of Springer Nature 2020

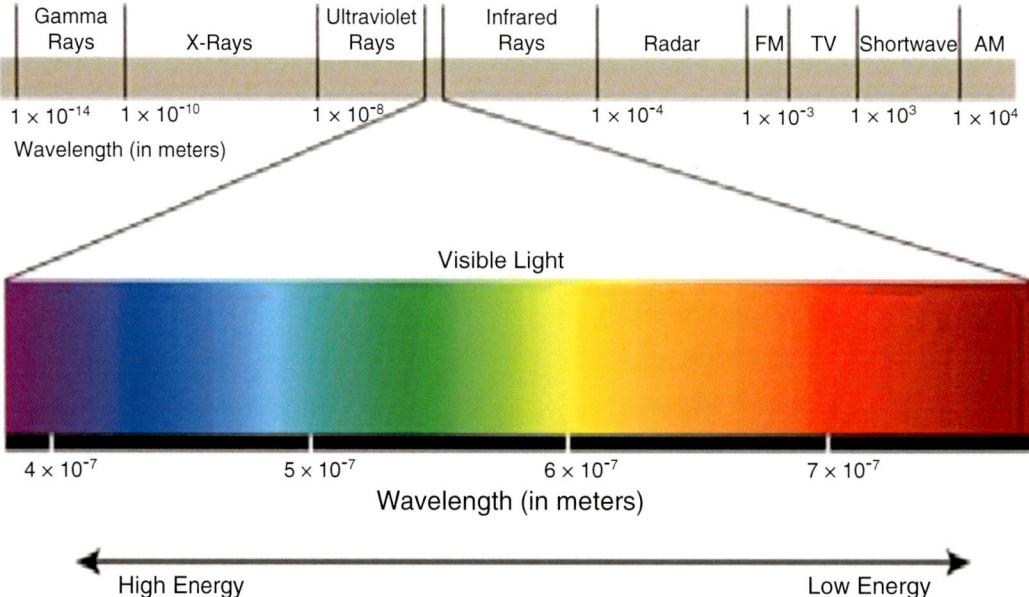

Fig. 1 Spectra of light

All the frequencies of white light (all colors) are absorbed by a black-colored car while all the frequencies of white light (all colors) are reflected by a white-colored car. So, a black-colored car is hotter than a white-colored car during summer. As we move from left to right side, the frequency (vibrations/sec) of light decreases and wavelength (length of a wave) increases as shown in Fig. 1.

Sometimes we get a line of colors so they are called as line spectra or discontinued spectra or emission spectra, e.g., hydrogen spectra.

- *Max Planck = Light has particle nature.*
- *Huygens = Light has wave nature.*
- *Einstein = Light has both (particle and wave) natures.*

As an atom, the smallest part of the light is called a photon. It can be imagined as an energy packet.

- Speed of light $= 3.0 \times 10^8$ m/s.
- RGB (red, green, and blue) are basic colors and all others are the derivatives of these primary colors.
- *Spectroscopy is dependent on emission while spectrophotometry is dependent on absorption of light.*
- In the initial days, some radiation was expressed as a mysterious thing which could not be seen but could be experienced. Hence, the word "spectra" comes from the Latin word "spectre," which means ghost.

The following are possibilities when the light is targeted to an object:

- Transmission (if the object is transparent)
- Absorption (if the object is liquid/gas or capable of absorption)
- Reflection (if an opaque material is placed)
- Refraction (if the medium is changed during the path of light)
- Scattering (if the surface is rough)

2 Colorimetry

- *Monochromatic light:* A single-colored light which has a particular frequency is called monochromatic light.
- *Monochromators:* Instruments used for getting a monochromatic light are called as monochromators, e.g., prism, grating, etc.

3 Beer's and Lambert's Law

As a monochromatic light is passed through a solution, some intensity is absorbed in the solution. So, a difference between the intensity of initial light and intensity of transmitted light occurs.

Beer's Law: *As the concentration increases, the absorption increases and so the intensity of transmitted light decreases exponentially as shown in Fig. 2.*

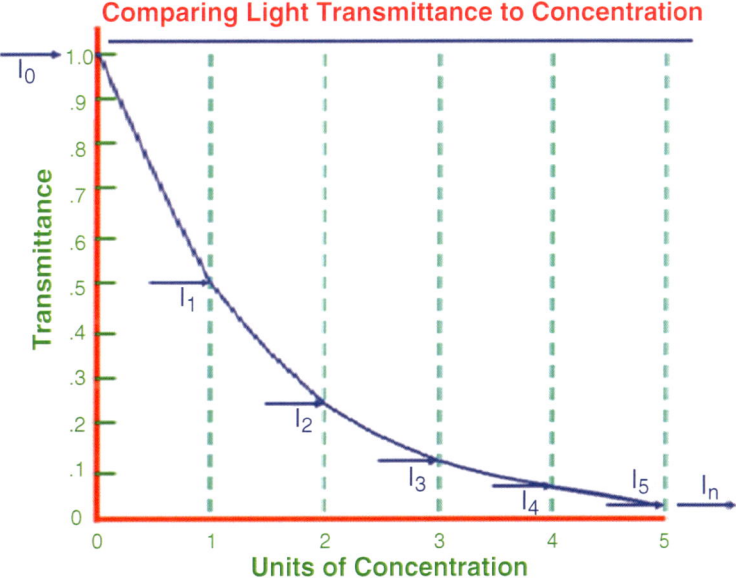

Fig. 2 The graph for Beer's law

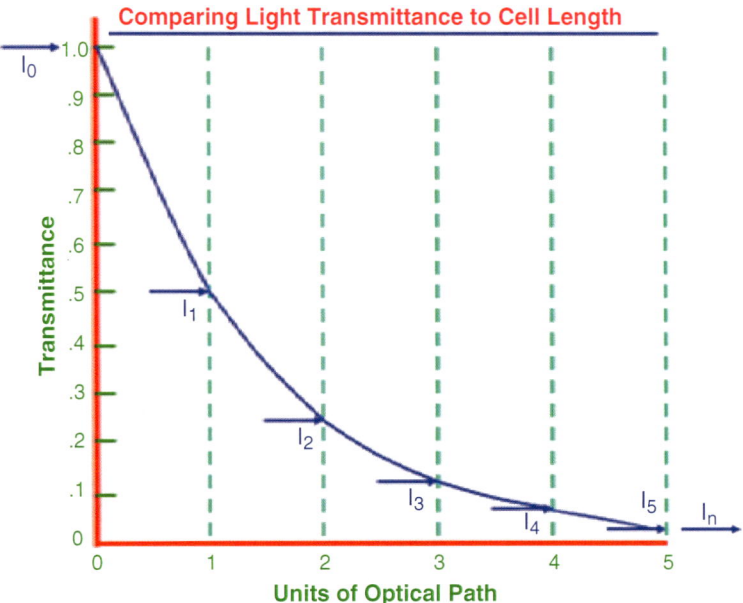

Fig. 3 Graph for Lambert's law

Lambert's Law: *As the thickness (or sometimes taken as cell length) increases, the absorption increases and so the intensity of transmitted light decreases exponentially as shown in Fig. 3.*

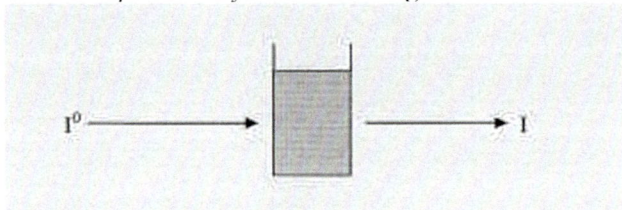

$$It = Io. \, e^{-kc} \quad It = Io. \, e^{-kT}$$

Here, Io = Intensity of initial (original) light

It = Intensity of transmitted light

k = Constant

C = Concentration

T = Thickness of the solution

As given in the above Fig. 4, as the transmission increases, absorbance decreases exponentially.

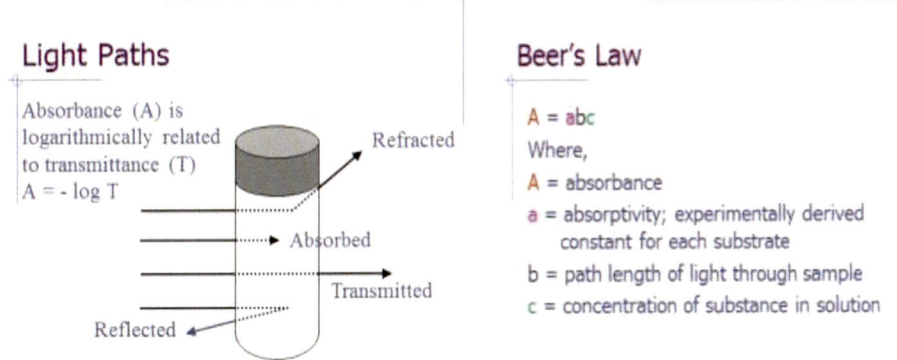

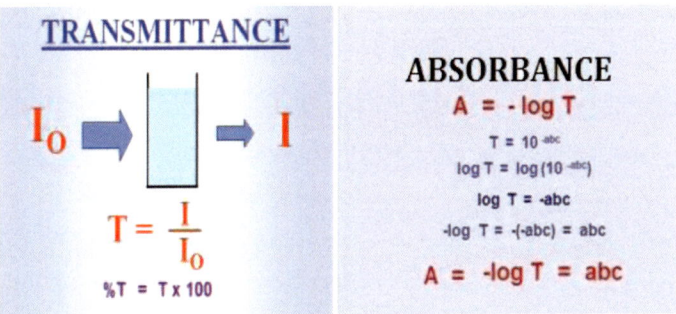

Fig. 4 The transmission of light

4   Deviation in Beer's and Lambert's Law

If a substance is following Beer's law, we get a straight line which is passing from the origin and the slope is "ab." But sometimes we do not get a straight line. If we are getting much more value of absorbance than desired, then deviation is said as positive deviation. But if we are getting much less value of absorbance than desired, then deviation is said as negative deviation as shown in Fig. 5.

There are two reasons for these deviations:

1. *Instrumental errors – The causes of instrumental errors are as follows:*

- Fluctuation in electricity.
- Light source is weak or malfunctioning.
- Arrangement of filter/monochromator is not proper.
- Scattering of light is happening inside the instrument.
- Slit is not placed properly.
- Outside knob is not working according to inner instrumentation.
- Sensitivity of detector is low or malfunctioning.

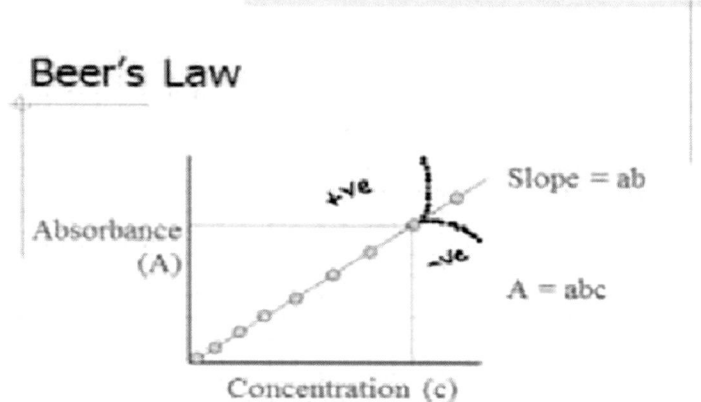

Fig. 5 Beer's law graph

2. *Chemical errors – The causes of chemical errors are as follows:*
   - Presence of bacteria.
   - Solution is cloudy.
   - Pigmentation is there in the solution.
   - Acid–base reaction is happening in the solution.
   - Association–dissociation reaction is happening in the solution.
   - Polarization reaction is happening in the solution.
   - If the color of the solution is changing with time.
   - If the absorption is happening due to solvent rather than solute.
   - This law is only applicable to some extent of concentration of the solution. Below or above that concentration, these laws are not applicable.

5   Instrumentation of Colorimeter

- Colorimeter is an instrument which compares the amount of light getting through an unknown solution and the amount of light getting through a pure solvent.

   The light is passed through a filter (to get the desired frequency of light) and concentrated using the lens. Then this light is targeted on a sample (which is kept in the cuvette). The light transmits through the sample and the transmitted light is measured by the detector. As the intensity of the initial light is known, we can find the difference between intensity of initial light and intensity of transmitted light as shown in Fig. 6.

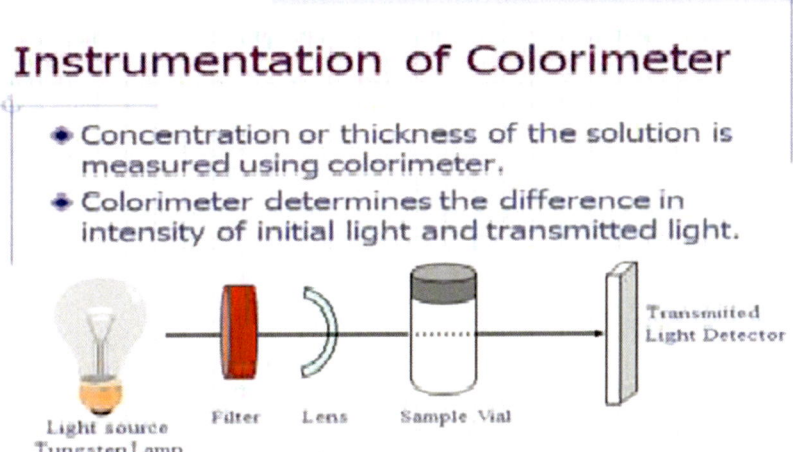

Instrumentation of Colorimeter

● Concentration or thickness of the solution is measured using colorimeter.

● Colorimeter determines the difference in intensity of initial light and transmitted light.

Light source Tungsten Lamp   Filter   Lens   Sample Vial   Transmitted Light Detector

Fig. 6 Instrument of colorimeter

Practical Application of Colorimeter:

- First of all, the instrument is set to 100% transmission (0% absorption) for cuvette with the solvent only.

- Then known concentrations of the desired solution are prepared, e.g., 1 ppm, 2 ppm, 3 ppm, etc.

- Then the absorptions related to the known concentrations are noted using the colorimeter.

- From the readings of absorption, a calibration curve is drawn (graph: absorption vs. concentration).

- Then the absorption of the known concentration of known solution is acquired using the colorimeter instrument, and from the calibration curve, the concentration of the unknown solution could be acquired (Figs. 7 and 8).

Prerequisite for a Solution to Be Analyzed by Colorimeter:

- The solution must be colored.

- The solution must not have any contaminant like bacteria, must not be cloudy, and must not have pigmentation. The solution must be clear (transparent).

- There must not be any reaction happening in the solution like acid–base, association–dissociation, or polarization reaction.

- The solution must have a particular concentration because Lambert's and

  Beer's laws can be applied only for a particular range of concentration.

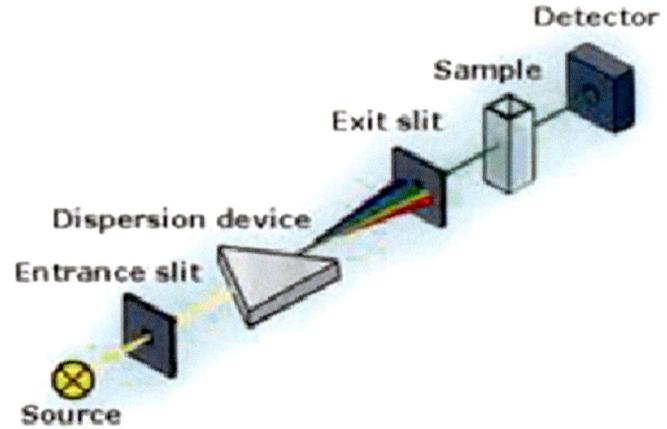

Fig. 7 Single-beam spectrophotometer

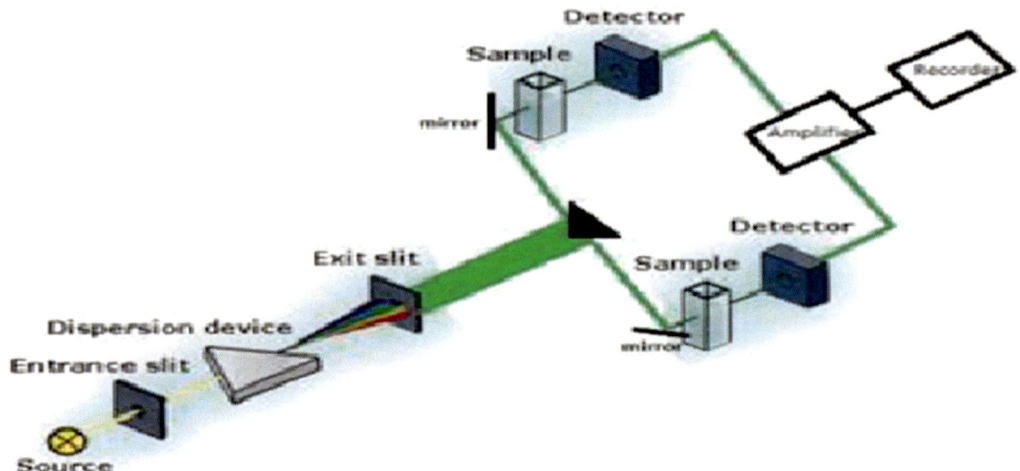

Fig. 8 Double-beam spectrophotometer

Difference Between Filter and Monochromators: *By filter, a specific band of frequency can be acquired, while by monochromators, a specific ray of light can be acquired (Table 1).*

Wavelength Selection Method:

1. Based on the color of the solution, another filter can be used, because the color of the solution is the color which is not absorbed by the solution. For example, in the case of a $CuSO_4$ solution, use a red-colored filter.

2. Find the highest absorption using different filters of different colors.

Table 1
Difference between (monochromators) prism and grating

S. no.	Prism	Grating
1.	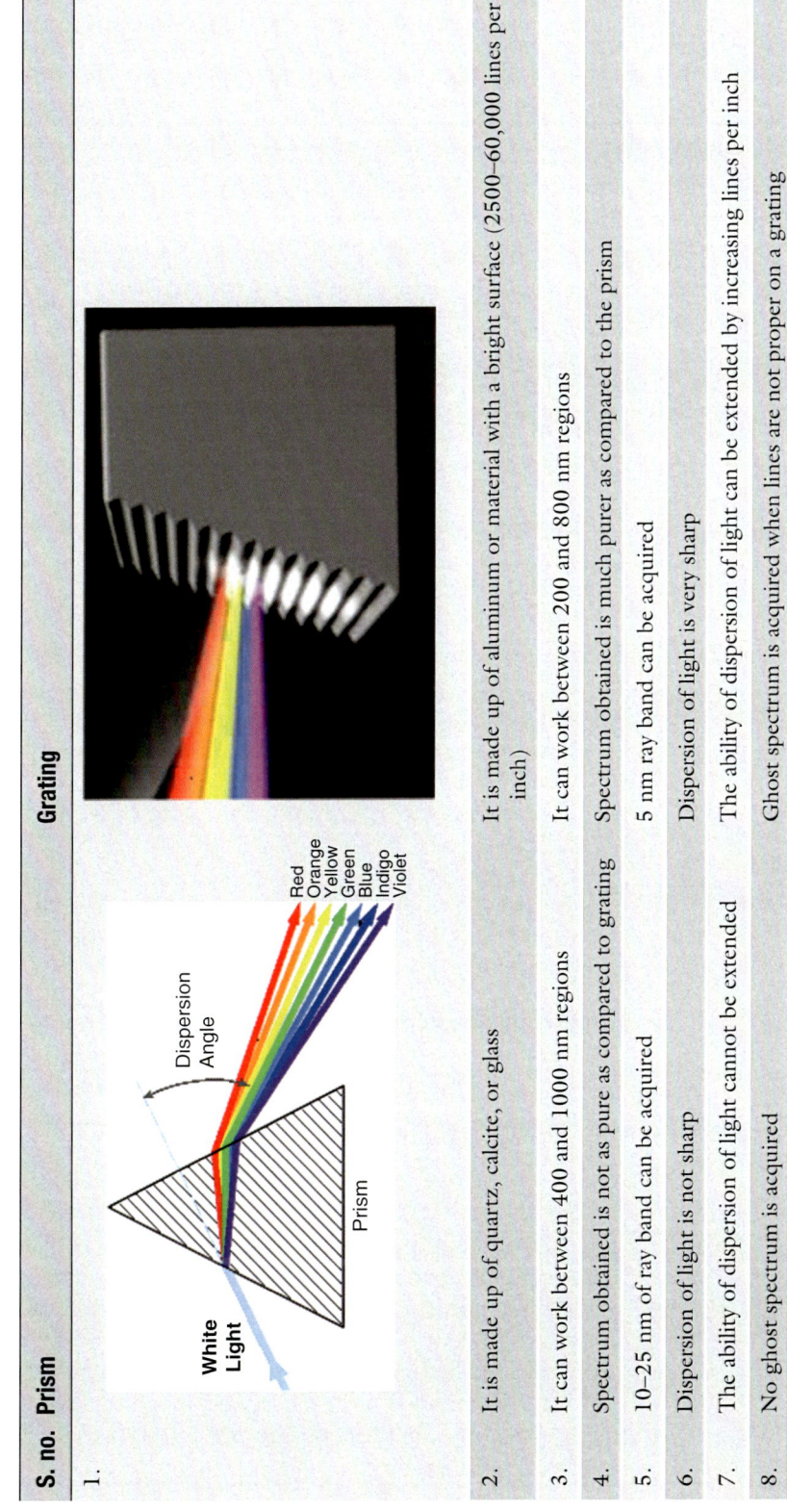	
2.	It is made up of quartz, calcite, or glass	It is made up of aluminum or material with a bright surface (2500–60,000 lines per inch)
3.	It can work between 400 and 1000 nm regions	It can work between 200 and 800 nm regions
4.	Spectrum obtained is not as pure as compared to grating	Spectrum obtained is much purer as compared to the prism
5.	10–25 nm of ray band can be acquired	5 nm ray band can be acquired
6.	Dispersion of light is not sharp	Dispersion of light is very sharp
7.	The ability of dispersion of light cannot be extended	The ability of dispersion of light can be extended by increasing lines per inch
8.	No ghost spectrum is acquired	Ghost spectrum is acquired when lines are not proper on a grating

3. The wavelength recommended for different solutions is given in the SOP (standard operating procedure) manual of the instrument. So use that filter.

---

6   Spectrophotometry

UV–Visible Spectroscopy

Principles

Quantification of Light Absorption: The chance for a photon to be absorbed by matter is given by an extinction coefficient which itself is dependent on the wavelength l of the photon. If light with the intensity I_0 passes through a sample with appropriate transparency and path length (thickness) d, the intensity I drops along the pathway in an exponential manner. The characteristic absorption parameter for the sample is the extinction coefficient a, yielding the correlation $I = I_0 e^{ad}$. The ratio $T = I/I_0$ is called transmission. Biochemical samples usually comprise aqueous solutions, where the substance of interest is present at a molar concentration c. Algebraic transformation of the exponential correlation into an expression based on the decadic logarithm yields the Beer–Lambert law.

The Beer–Lambert law is valid for low concentrations only. Higher concentrations might lead to association of molecules and therefore cause deviations from the ideal behavior. Absorbance and extinction coefficients are additive parameters, which complicate determination of concentrations in samples with more than one absorbing species. Note that in dispersive samples or suspensions scattering effects increase the absorbance, since the scattered light is not reaching the detector for read-out. The absorbance recorded by the spectrophotometer is thus overestimated and needs to be corrected.

Deviations from the Beer–Lambert Law: According to the Beer–Lambert law, absorbance is linearly proportional to the concentration of chromophores. This might not be the case anymore in samples with high absorbance. Every spectrophotometer has a certain amount of stray light, which is light received at the detector but not anticipated in the spectral band isolated by the monochromator. In order to obtain reasonable signal-to-noise ratios, the intensity of light at the chosen wavelength (I_l) should be ten times higher than the intensity of the stray light (I_{stray}). If the stray light gains in intensity, the effects measured at the detector have nothing or little to do with chromophore concentration. Secondly, molecular events might lead to deviations from the Beer–Lambert law. For instance, chromophores might dimerize at high concentrations and, as a result, might possess different spectroscopic parameters.

Instrumentation of Spectrophotometer: *There are two types of spectrophotometer available:*

1. *Single-beam spectrophotometer*
2. *Double-beam spectrophotometer*

A double-beam spectrophotometer splits the beam of light into two different paths, one of which passes through the sample while the other passes through a reference standard. However, single-beam spectrophotometers are usually more compact and have a higher dynamic range. UV–Vis spectrophotometers are usually double-beam spectrophotometers where the first channel contains the sample and the second channel holds the control (buffer) for correction.

Alternatively, one can record the control spectrum first and use this as internal reference for the sample spectrum. The latter approach has become very popular as many spectrophotometers in the laboratories are computer-controlled, and baseline correction can be carried out using the software by simply subtracting the control from the sample spectrum [1, 2].

The light source is a tungsten filament bulb for the visible part of the spectrum and a deuterium bulb for the UV region. Since the emitted light consists of many different wavelengths, a monochromator, consisting of either a prism or a rotating metal grid of high precision called grating, is placed between the light source and the sample. Wavelength selection can also be achieved by using colored filters as monochromators that absorb all but a certain limited range of wavelengths. This limited range is called the bandwidth of the filter. Filter-based wavelength selection is used in colorimetry, a method with moderate accuracy, but best suited for specific colorimetric assays where only certain wavelengths are of interest. If wavelengths are selected by prisms or gratings, the technique is called spectrophotometry [1, 2].

In a double-beam instrument, the incoming light beam is split into two parts by a half-mirror. One beam passes through the sample, the other through a control (blank, reference). This approach obviates any problems of variation in light intensity, as both reference and sample would be affected equally. The measured absorbance is the difference between the two transmitted beams of light recorded. Depending on the instrument, a second detector measures the intensity of the incoming beam, although some instruments use an arrangement where one detector measures the incoming and the transmitted intensity alternately. The latter design is better from an analytical point of view as it eliminates potential variations between the two detectors. At about 350 nm most instruments require a change of the light source from visible to UV light. This is achieved by mechanically moving mirrors that direct the appropriate beam along the optical axis and divert the

Table 2
For the different regions of light, these lamps (light sources) can be used

	EM region	Light source	Wavelength
1.	UV/near-IR region	Hydrogen or deuterium discharge lamp	10–200 nm
2.	Visible region	Tungsten lamp	200–1000 nm
3.	IR region	Nernst glower	1000–1,000,000 nm

other. When scanning the interval of 500–210 nm, this frequently gives rise to an offset of the spectrum at the switchover point.

Light Source: The light source must fulfill the following prerequisites:

The light coming from the source must have proper intensity.

The light source must have all the frequencies of light so that the required frequency can be acquired.

The light source must be stable. It must not change with time (Table 2).

7   Filter and Monochromators

From this part, the radiation with only specific wavelength could be passed. Other radiations are absorbed.

Filters: Filters allow only a small section of frequency to pass through and all others are absorbed as shown in Fig. 9.
  *Filters are used mostly in the colorimeter. They are made up of glass or gelatin.

Monochromators: *Optical devices used for selecting a specific wavelength from a range of frequency are called monochromators. They are of two types: slit and dispersive element.*
  For dispersion of light, prism or grating or both can be used (Table 3).

Sample vessel (cuvette): *In all the spectrophotometric analysis, the absorption of light is measured. So the cell used in the analysis must not absorb the light or the absorption must be minimum. Hence, for a different region of a spectrum as shown in Fig. 10,*
  different materials can be used as follows:

  1. *For UV range: Quartz cuvette*

  2. *For visible range: Glass cuvette*

  3. *For IR range: NaCl, KBr, and Nujol cuvette*

Detector: *The light which is initiated from the source and passed*

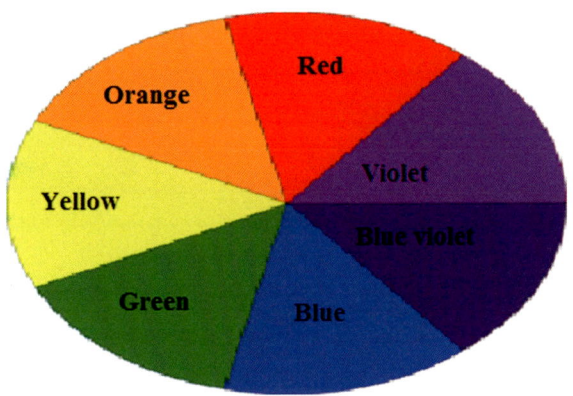

Fig. 9 Filters

Table 3
Different filters used to separate different regions of visible light

	The color of the visible region	Wavelength
1.	Violet	380–470 nm
2.	Blue	440–490 nm
3.	Blue (slightly greenish)	490–500 nm
4.	Green	500–560 nm
5.	Yellow (slightly greenish)	560–580 nm
6.	Yellow	580–600 nm
7.	Orange	600–650 nm
8.	Red	650–750 nm

Fig. 10 Cuvette

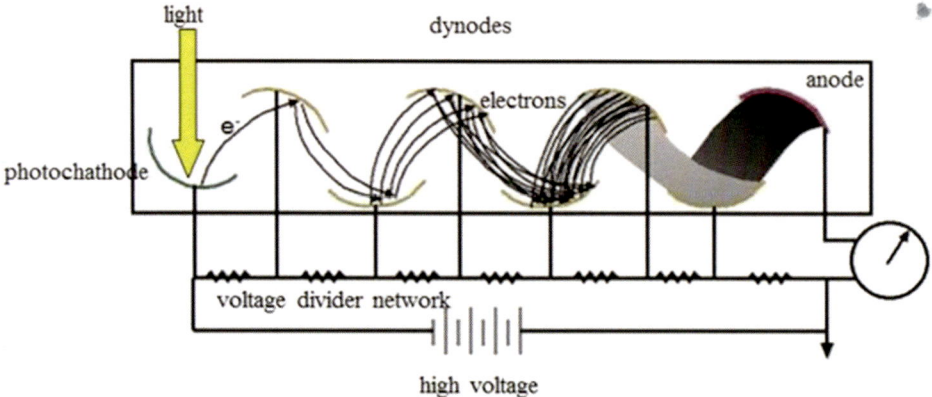

Fig. 11 The photomultiplier

through the cuvette is measured by the detector. Absorption or transmission of the light can be measured by the detector. Mainly three types of detectors can be used:

1. *Photovoltaic cell (barrier layer cell)*
2. *Phototubes (photoemissive tube)*
3. *Photomultiplier tubes*

Recorder: *The measurement of absorption or transmission is recorded in digital form in the recorder. The graph of wavelength vs. absorption can also be acquired. The photomultiplier is shown in Fig. 11.*

Photomultiplier Tube: * Due to very high amplification in photomultiplier tube, it is much sensitive than simple photocell tube as shown in Fig. 12. So it is very much used in a spectrophotometer.

There is a photocathode, the surface of which is coated with a light-reflecting material. There are also symmetrically arranged poles which are called dynodes. Each dynode also has light-reflecting material, which has more potential than sequential cathode. Dynode is set to +ve voltage. So, when light indented on the surface of the cathode, a primary electron is generated. These electrons flow to the neighboring dynode whose potential is 50–90 V more than the cathode. Moreover, each electron generates four to five secondary electrons. This process happens on almost nine dynodes. Thus, amplification happens in the tube and shower of the electron ($4^9 = 2.6 \times 10^6$) occurs. When this shower of the electron is captured by the detector, intensity of the light can be known and thus we can have a strong signal. To start the tube, 500–900 V of electricity is necessary which is applied using a number of batteries in a sequence.

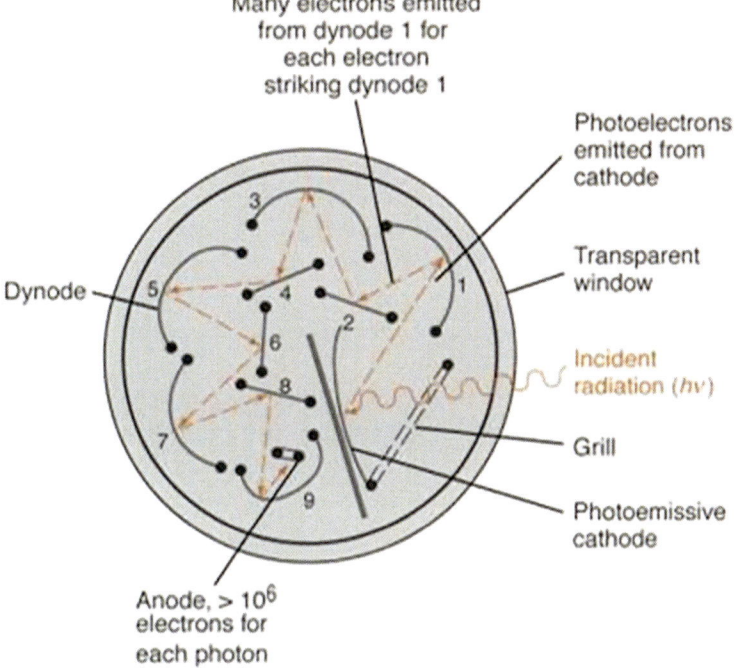

Many electrons emitted
from dynode 1 for
each electron
striking dynode 1

Photoelectrons
emitted from
cathode

Transparent
window

Dynode

Incident
radiation ($h\nu$)

Grill

Photoemissive
cathode

Anode, > 10^6
electrons for
each photon

Fig. 12 Photomultiplier tube

8   Applications

The usual procedure for (colorimetric) assays is to prepare a set of standards and produce a plot of concentration versus absorbance called calibration curve. This should be linear as long as the Beer–Lambert law applies. Absorbances of unknowns are then measured and their concentrations interpolated from the linear region of the plot. It is important that one never extrapolates beyond the region for which an instrument has been calibrated as this potentially introduces enormous errors.

To obtain good spectra, the maximum absorbance should be approximately 0.5 which corresponds to concentrations of about 50 mM (assuming e = 10,000 $dm^3 mol^1 cm^1$).

Qualitative and Quantitative Analysis: Qualitative analysis may be performed in the UV–Vis regions to identify certain classes of compounds both in the pure state and in biological mixtures (e.g., protein-bound). The application of UV–Vis spectroscopy to further analytical purposes is rather limited, but possible for systems where appropriate features and parameters are known.

Most commonly, this type of spectroscopy is used for quantification of biological samples either directly or via colorimetric assays. In many cases, proteins can be quantified directly using their intrinsic chromophores, tyrosine and tryptophan. Protein spectra are acquired by scanning from 500 to 210 nm. The characteristic features in a protein spectrum are a band at 278/280 nm and another at 190 nm. The region from 500 to 300 nm provides valuable information about the presence of any prosthetic groups or coenzymes. Protein quantification by single wavelength measurements at 280 and 260 nm only should be avoided, as the presence of larger aggregates (contaminations or protein aggregates) gives rise to considerable Rayleigh scatter that needs to be corrected.

Common applications for difference UV spectroscopy include the determination of the number of aromatic amino acids exposed to solvent, detection of conformational changes occurring in proteins, detection of aromatic amino acids in active sites of enzymes, and monitoring of reactions involving "catalytic" chromophores (prosthetic groups, coenzymes).

References

1. Rajan K (2011) Analytical techniques in biochemistry and molecular biology. Springer, New York. eBook ISBN978-1-4419-9785-2

2. Wilson K, Walker J (2010) Principles and techniques of biochemistry and molecular biology. Cambridge University Press, Cambridge. ISBN 978-0-521-73167-6

Chapter 5

Chromatography

Abstract

The successful chromatographic separation of analytes in a mixture depends upon the selection of the most appropriate process of chromatography followed by the optimization of the experimental conditions associated with the separation. Optimization requires an understanding of the processes that are occurring during the development and elution and of the calculation of a number of experimental parameters characterizing the behavior of each analyte in the mixture.

In any chromatographic separation, two processes occur concurrently to affect the behavior of each analyte and hence the success of the separation of the analytes from each other. The first involves the basic mechanisms defining the chromatographic process such as adsorption, partition, ion exchange, ion pairing, and molecular exclusion. These mechanisms involve the unique kinetic and thermodynamic processes that characterize the interaction of each analyte with the stationary phase. The second general process defines the other processes, such as diffusion, which tend to oppose the separation and which result in non-ideal behavior of each analyte. This chapter explains about paper and thin-layer chromatography, HPLC, GC, and gel filtration chromatography.

Key words Paper chromatography, Thin-layer chromatography, HPLC, GC, Gel filtration chromatography

Abbreviations

HPLC   High-performance liquid chromatography
RF   Relative flow
TLC   Thin-layer chromatography

---

1   Introduction

History: *Mikhail Tsvet (1872–1919), a Russian botanist, in 1906 used chromatography to separate plant pigments. He called the new technique chromatography because the result of the analysis was "written in color" along the length of the adsorbent column. Chroma means "color" and graphein means to "write" [1, 2].*

Importance: *Chromatography has application in every branch of the physical and biological sciences. Twelve Nobel Prizes were awarded*

Mahin Basha, *Analytical Techniques in Biochemistry*, Springer Protocols Handbooks,
https://doi.org/10.1007/978-1-0716-0134-1_5, © Springer Science+Business Media, LLC, part of Springer Nature 2020

between 1937 and 1972 alone for work in which chromatography played a vital role.

Chromatography is a method of separating a mixture of molecules depending on their distribution between a mobile phase and a stationary phase. The mobile phase (also known as solvent) may be either liquid or gas. The stationary phase (also known as sorbent) can be either a solid or liquid; a liquid stationary phase is held stationary by a solid. The solid holding the liquid stationary phase is the support or matrix.

2   Types of Chromatography

- Column chromatography
- Ion-exchange chromatography
- Gel-permeation (molecular sieve) chromatography
- Affinity chromatography
- Paper chromatography
- Thin-layer chromatography
- Gas chromatography
- Dye-ligand chromatography
- Hydrophobic interaction chromatography
- Pseudoaffinity chromatography
- High-pressure liquid chromatography (HPLC)

3   Partition Chromatography

The distribution of solutes between two immiscible phases. The solute will distribute itself between the two phases according to its solubility in each phase; this is called partitioning.

The two most common types of partition chromatography are thin-layer chromatography and paper chromatography. In both cases, the stationary phase is a liquid bound to a matrix. In paper chromatography, the stationary phase is water molecules bound to a cellulose matrix.

Distribution Coefficients: *The basis of all forms of chromatography is the distribution or partition coefficient (Kd), which describes the way in which a compound (the analyte) distributes between two immiscible phases. For two such phases A and B, the value for this coefficient is a constant at a given temperature and is given by the expression:*

$$\frac{\text{Concentration in phase } \mathbf{A}}{\text{Concentration in phase } \mathbf{B}} ^{1/4} \mathbf{K_d}$$

3.1 Paper Chromatography

- The cellulose support contains a large amount of bound water.
- Partitioning occurs between the bound water which is the stationary phase and the solvent which is the mobile phase.

Principles of Paper Chromatography:

Capillary action – The movement of liquid within the spaces of a porous material due to the forces of adhesion, cohesion, and surface tension. The liquid is able to move up the filter paper because its attraction to itself is stronger than the force of gravity.

Solubility – The degree to which a material (solute) dissolves into a solvent. Solutes dissolve into solvents that have similar properties (like dissolves like). This allows different solutes to be separated by different combinations of solvents.

Separation of components depends on both their solubility in the mobile phase and their differential affinity to the mobile phase and the stationary phase.

Experimental Procedure for Paper Chromatography:

- A small volume of a solution or a mixture to be separated or identified is placed at a marked spot (origin) on a sheet or strip of paper and allowed to dry.
- The paper is then placed in a closed chamber and one end is immersed in a suitable solvent.
- The solvent is drawn (moved) through the paper by capillary action.
- As the solvent passes the origin, it dissolves the sample and moves the components in the direction of flow.
- After the solvent front has reached a point near the other end of the paper, the sheet or strip is removed and dried.
- The spots are then detected and their positions marked as shown in Fig. 1.
- The ratio of the distance moved by a solute to the distance moved by the solvent $=$ R_f.
- The R_f is always less than one.

The sample is deposited at the bottom line of the paper. Paper is placed in a tank filled with 1 cm solvent. Solvent migrates in the paper and elutes the solutes. The solutes migrate depending on their affinity for the solvent. Techniques of development with various flow directions are shown in Fig. 2.

Chromatogram:

- Once a sample is applied on TLC or paper, it is called chromatogram.
- Paper chromatogram can be developed either by ascending or descending solvent flow.

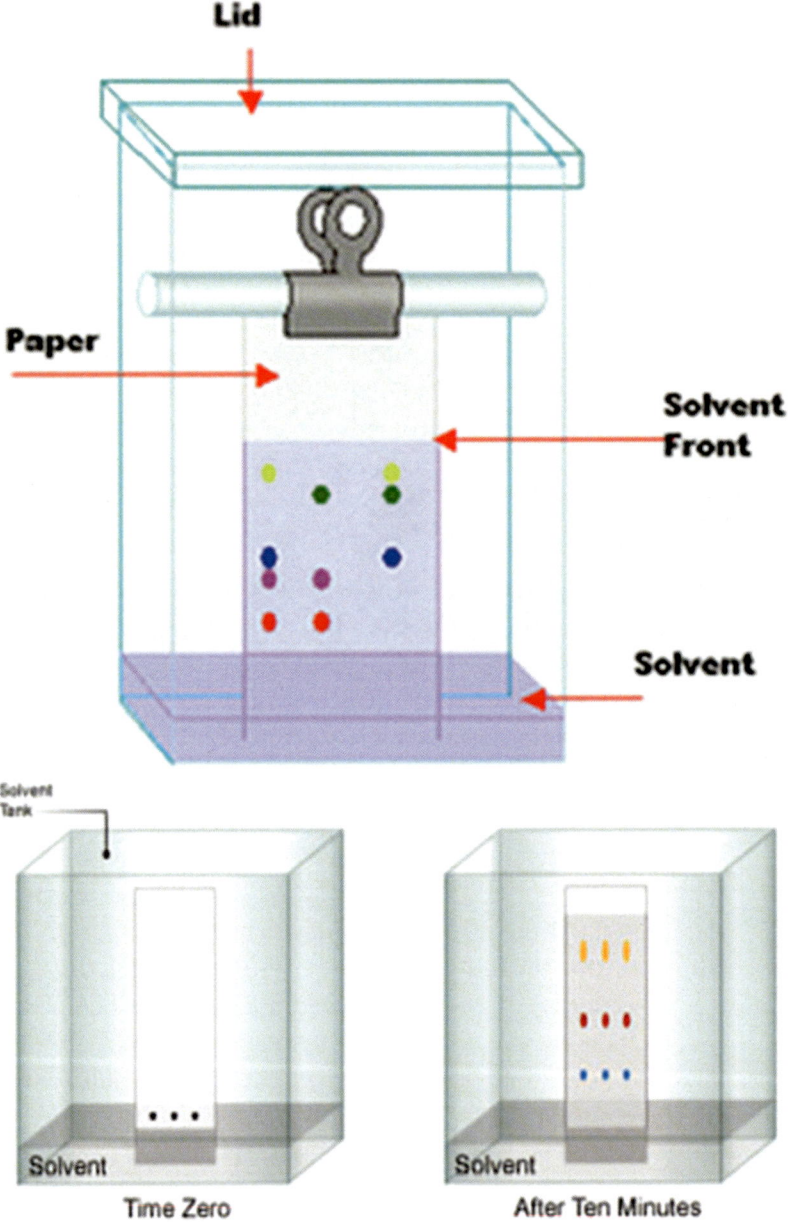

Fig. 1 Paper chromatography

- Descending chromatography is faster because gravity helps the solvent flow.
- Disadvantages: It is difficult to set the apparatus.
- Ascending is simple and inexpensive compared with descending and usually gives more uniform migration with less diffusion of the sample "spots."

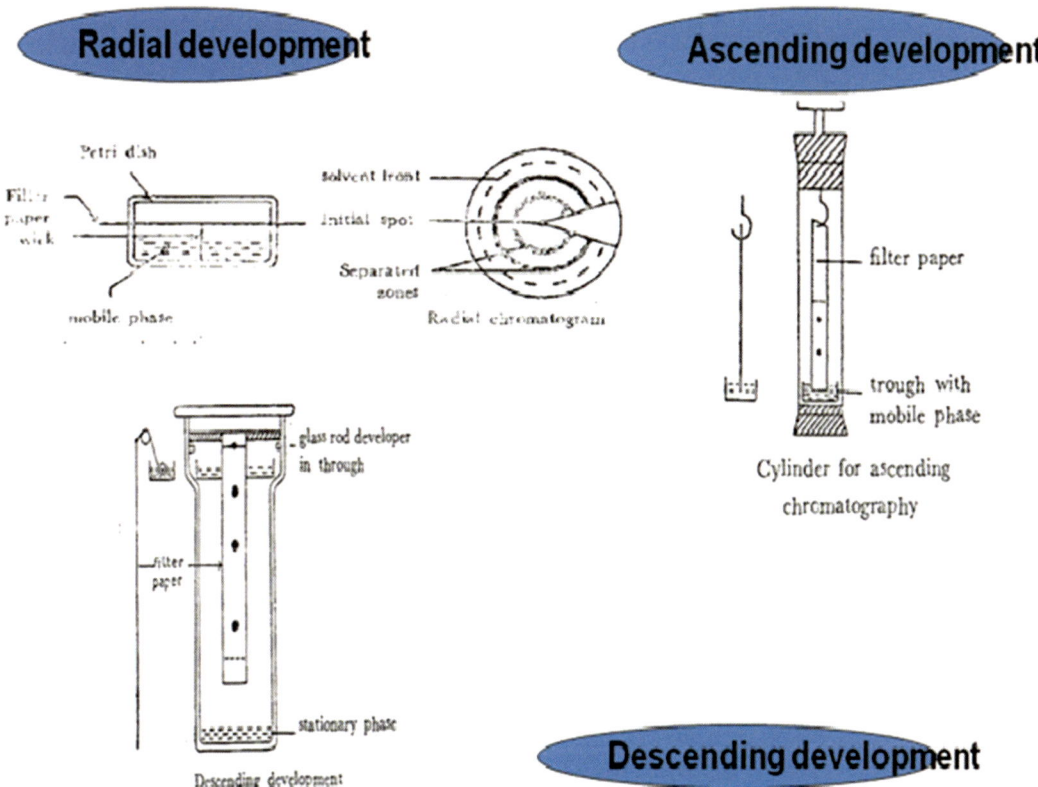

Fig. 2 Techniques of development with various flow directions

Detection of Spots:

- Spots in paper chromatograms can be detected in four different ways:

    1. By their natural color.
    2. By their fluorescence.
    3. By their chemical reactions that take place after the paper has been sprayed with various reagents; for example, during paper chromatography of amino acids, the chromatograms are sprayed with ninhydrin.
    4. By radioactivity.

    The spots are usually identified by comparing standards of known Rf values.

3.2 Thin-Layer Chromatography

- In TLC, the stationary phase is the solvent added to the support to form the thin layer so the solvent gets bound to the matrix (support).
- Partition chromatography is mainly used for separation of molecules of small molecular weight.

Paper chromatography uses paper which can be prepared from cellulose products only. In TLC, any substance that can be finely divided and formed into a uniform layer can be used. Both organic and inorganic substances can be used to form a uniform layer for TLC. Organic substances include cellulose, polyamide, and polyethylene. Inorganic substances include silica gel, aluminum oxide, and magnesium silicate.

It is a method for identifying substances and testing the purity of compounds. TLC is a useful technique because it is relatively quick and requires small quantities of minerals.

Advantages of TLC over Paper Chromatography:

- Greater resolving power because there is less diffusion of spots
- Greater speed of separation
- Wide choice of materials as sorbents

Preparing the Chamber:
To a jar with a tight-fitting lid, add enough of the appropriate developing liquid so that it is 0.5–1 cm deep in the bottom of the jar as shown in Fig. 3. Close the jar tightly, and let it stand for about 30 min so that the atmosphere in the jar becomes saturated with solvent. Partition of a solute between a moving solvent phase and a stationary aqueous phase: The solute moves in the direction of a solvent flow at a rate determined by the solubility of the solute in the moving phase. Thus a compound with high mobility is more attracted to the moving organic phase than to the stationary phase.

Ion exchange effect: Any ionized impurities in the support medium will tend to bind or attract oppositely charged ions (solutes) and will, therefore, reduce the mobility of these solutes. Temperature: Since temperature can affect the solubility of the solute in a given solvent, temperature is also an important factor.

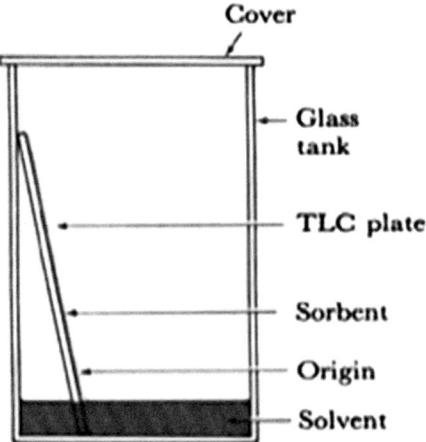

Fig. 3 Thin-layer chromatography (TLC) instrument

The molecular weight of a solute also affects the solubility and hence chromatographic performance (Figs. 4, 5, and 6).

Adsorption of compound (solute) onto support medium: Although the support medium (silica gel) is theoretically inert,

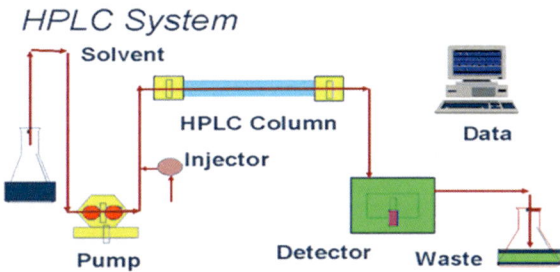

Fig. 4 Schematic diagram of HPLC

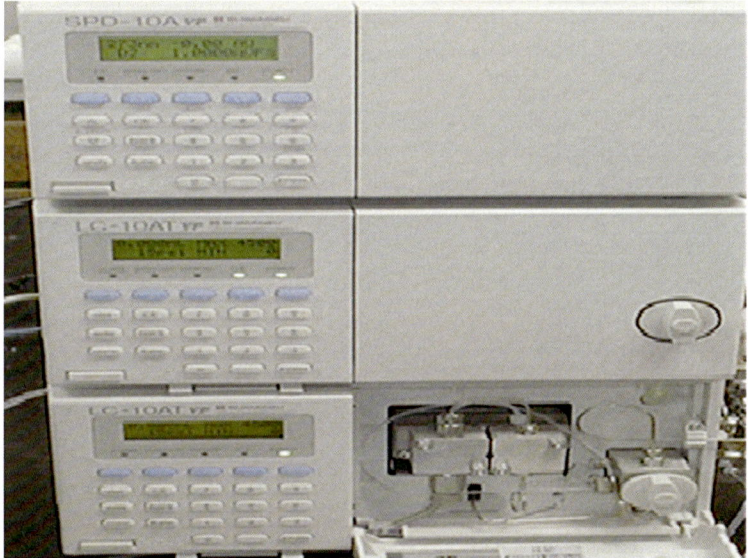

Fig. 5 Picture of HPLC instrument

Fig. 6 Picture of an HPLC column

this is not always the case. If a solute tends to bind to the support medium, this will slow down its mobility in the solvent system. The composition of the solvent: Since some compounds are more soluble in one solvent than in the other, the mixture of solvents used will affect the separation of compounds.

Expression of the Results:

- The term "*Rf*" (relative flow) is used to express the performance of a solute in a given solvent system/support medium. The term *Rf* value may be defined as the ratio of the distance the compound migrates to the distance the solvent migrates. *Rf* value is constant for a particular compound, solvent system, and insoluble matrix.

$$Rf = \frac{\text{Distance of migration of solute}}{\text{Distance moved by solvent}}$$

4   Uses for Chromatography

Chromatography is used by scientists to:

- *Analyze* – Examine a mixture, its components, and their relations to one another
- *Identify* – Determine the identity of a mixture or components based on known components
- *Purify* – Separate components in order to isolate one of interest for further study
- *Quantify* – Determine the amount of the mixture and/or the components present in the sample

Real-life examples of uses for chromatography:

- *Pharmaceutical company* – Determine the amount of each chemical found in new product
- *Hospital* – Detect blood or alcohol levels in a patient's bloodstream
- *Law enforcement* – Compare a sample found at a crime scene to samples from suspects
- *Environmental agency* – Determine the level of pollutants in the water supply
- *Manufacturing plant* – Purify a chemical needed to make a product

5 Column Chromatography

In column chromatography, the stationary phase is packed into a glass or metal column. The mixture of analytes is then applied and the mobile phase, commonly referred to as the eluent, is passed through the column either by use of a pumping system or applied gas pressure. The stationary phase is either coated onto discrete small particles (the matrix) and packed into the column or applied as a thin film to the inside wall of the column. As the eluent flows through the column, the analytes separate on the basis of their distribution coefficients and emerge individually in the eluate as it leaves the column.

The two forms of column chromatography to be discussed in this chapter are liquid chromatography (LC), mainly high-performance liquid chromatography (HPLC), and gas chromatography (GC).

5.1 HPLC

- It is also called as high-pressure liquid chromatography or high-performance liquid chromatography. *Stationary phase* may be solid (adsorption) or liquid (partition). *Mobile phase* may be gas (GC) or liquid (LC) .It is a technique by which a mixture sample is separated into components for identification, quantification, and purification of mixtures.

Retention Time:

A chromatogram is a pictorial record of the detector response as a function of elution volume *or retention time*. It consists of a series of peaks or bands, *ideally symmetrical* in shape, representing the elution of individual analytes. The retention time t_R for each analyte has two components. The first is the time it takes the analyte molecules to pass through the free spaces between the particles of the matrix coated with the stationary phase. This time is referred to as the dead time, t_M. The volume of the free space is referred to as the column void volume, V_0.

Retention Factor:

One of the most important parameters in chromatography is the retention factor, k (previously called capacity factor and represented by the symbol k^0). It is simply the additional time that the analyte takes to elute from the column relative to an unretained or excluded analyte that does not interact with the stationary phase and which, by definition, has a k value of 0.

Instrumentation:

Principle

Partitioning: Separation is based on the analyte's relative solubility between two liquid phases.

HPLC – Modes
- Normal phase – Polar stationary phase and a non-polar solvent
- Reverse phase – Non-polar stationary phase and a polar solvent.

Common Solvents of Reverse Phase:

Methanol – Most common solvent, close to water in structure. Miscible in all proportions with H_2O. So, for less polar organics, you can have the power of 90% methanol with 10% H_2O.

Acetonitrile – Highly polar, very low UV absorbance. Also completely miscible with H_2O but *lacking in hydrogen bonding* capability thus affording a different partitioning effect.

Tetrahydrofuran – Molecule has a high dipole moment. More soluble with non-polar compounds.

Water – Also a very common solvent. Used to make up solvent modifiers to adjust pH (buffers) as well as ion-pairing reagents.

The heart of an HPLC system is the *column*. The column contains the particles that contain the stationary phase. The mobile phase is pumped through the column by *a pump*. Solvents must be *degassed* to eliminate the formation of bubbles. Guard column is to protect the main column from impurities. Impurities get filtered and pure sample reaches the main column.

1. *Pump*

The role of the pump is to force a liquid (mobile phase) through the liquid chromatography at a specific flow rate. A pump can deliver a constant mobile phase composition in which (isocratic) the flow per hour (f.ph) composition remains unchanged during the analysis or (gradient) the f.ph changes during the analysis.

2. *Injector*
- The injector serves to introduce the liquid sample into the flow stream of the mobile phase. It may be auto-sampler or manual.

3. *There are a wide variety of stationary phases available for HPLC:*
   Normal phase: Polar stationary phase and a non-polar solvent, e.g., silica gel

   Reverse phase: Non-polar stationary phase and a polar solvent, e.g., silica gel C18

   Ion Exchange: Stationary phase contains ionic groups and the mobile phase is an aqueous buffer.

   Size Exclusion: There is no interaction between the sample compounds and the column. Large molecules elute first. Smaller molecules elute later as shown in Fig. 7.

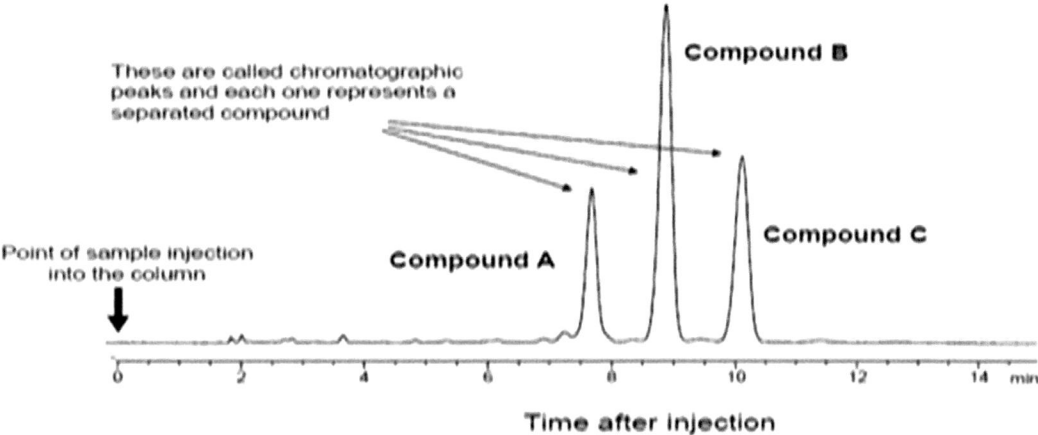

Fig. 7 Chromatogram

6   Parameters of HPLC

Qualitative Analysis:

The most common parameter for a compound is retention time (the time it takes for that specific compound to elute from the column after injection) as shown in Fig. 8.

Capacity factor (k'):

Is a measure for the position of a sample peak in the chromatogram

$$k' = (t_{R1} - t_o)/t_o$$

Selectivity factor (a):

Also called separation or selectivity coefficient is defined as

$$\alpha = k_2'/k_1' = (t_{R2} - t_o)/(t_{R1} - t_o)$$

Quantitative Analysis:

The measurement of the amount of compounds in a sample (concentration) as shown in Fig. 9.

1. Determination of the peak height
2. Determination of the peak area

Resolution (R_S) of a column provides a quantitative measure of its ability to separate two analytes.

$$R_S = 2(TR2 - TR1)/W2 + W1$$

Theoretical plates (N): The number of theoretical plates characterizes the efficiency of a column.

$$N = 16(t_R/W)^2$$

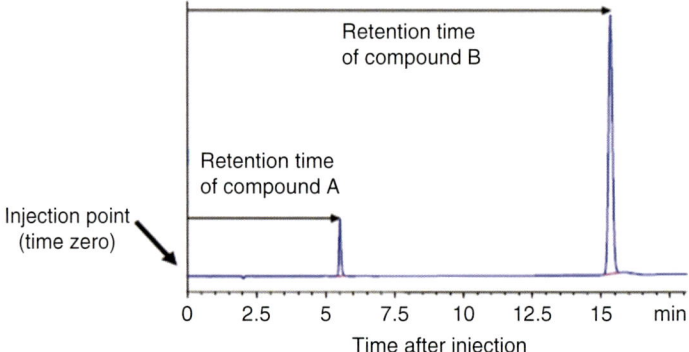

Fig. 8 Retention time of a compound to elute from the column after injection

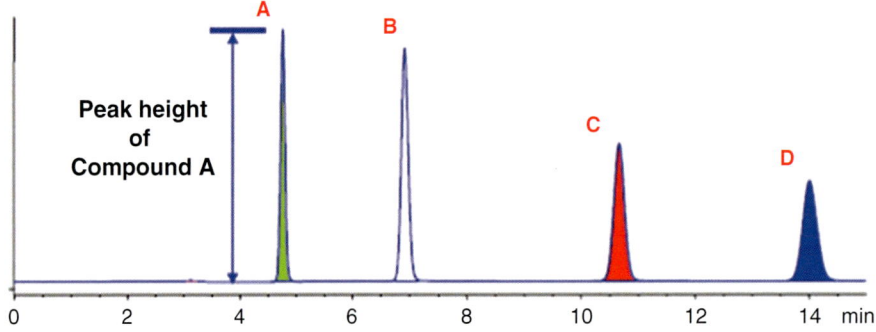

Fig. 9 Peak height of acompound

Evaluation of HPLC Parameters:

- Efficiency
- Resolution
- Inertness
- Retention index
- Column bleed
- Capacity factor

Uses of HPLC:

This technique is used in chemistry and biochemistry research for analyzing complex mixtures, purifying chemical compounds, developing processes for synthesizing chemical compounds, isolating natural products, or predicting physical properties. It is also used in quality control to ensure the purity of raw materials, to control and improve process yields, to quantify assays of final products, or to evaluate product stability and monitor degradation.

In addition, it is used for analyzing air and water pollutants, for monitoring materials that may jeopardize occupational safety or

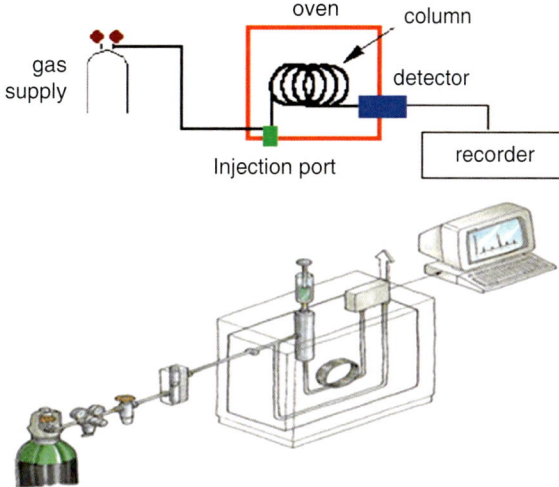

Fig. 10 Gas cromatography instrumentation

health, and for monitoring pesticide levels in the environment. Federal and state regulatory agencies use HPLC to survey food and drug products, to identify confiscated narcotics, or to check for adherence to label claims.

6.1 Gas Chromatography

- Gas chromatography is a chromatographic technique that can be used to separate *volatile organic compounds*. This method depends upon the solubility and boiling points of organic liquids in order to separate them from a mixture. It is both a qualitative (identity) and quantitative (how much of each) tool as shown in Fig. 10.

  It consists of:

- A flowing mobile phase
- An injection port
- A separation column (the stationary phase)
- An oven
- A detector

7 Instrumentation

Principle: *The organic compounds are separated due to differences in their* partitioning behavior *between the mobile gas phase and the stationary phase in the column as shown in Fig. 11.*

- *Mobile phases are generally inert gases such as helium, argon, or* nitrogen.

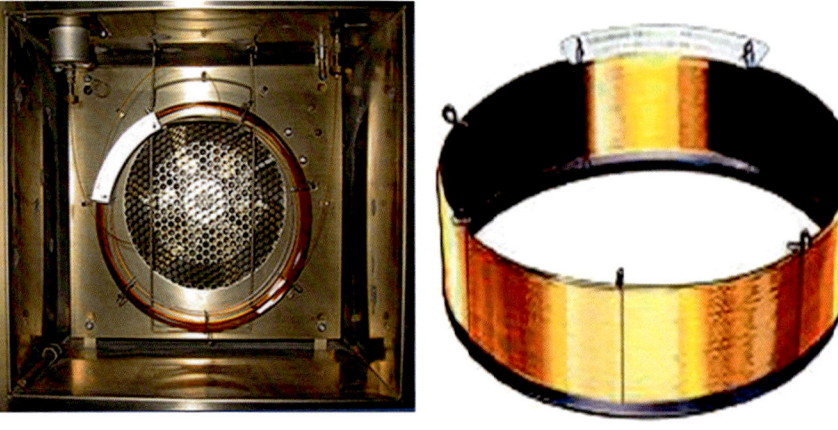

Capillary column

Fig. 11 A gas chromatography oven, open to show a capillary column

- *The injection port consists of a rubber septum through which a syringe needle is inserted to inject the sample.*

- *The injection port is maintained at a higher temperature than the boiling point of the least volatile component in the sample mixture.*

- *Since the partitioning behavior is dependent on* temperature, *the separation column is usually contained in a thermostat-controlled oven.*

- *Separating components with a wide range of boiling points is accomplished by starting at a low oven temperature and increasing the temperature over time to elute the high-boiling-point components.*

GC Columns

Packed columns	Capillary columns
•Typically a glass or stainless steel coil. •1-5 total length and 5 mm inner diameter. • Filled with the st. ph. or a packing coated with the st.ph.	•Thin fused-silica. •Typically 10-100 m in length and 250 μm inner diameter. •St. ph. coated on the inner surface. •Provide much higher separation eff. •But more easily overloaded by too much sample.

GC Detectors:
After the components of a mixture are separated using gas chromatography, they must be detected as they exit the GC column. Thermal-conductivity (TCD) and flame ionization (FID) detectors are the two most common detectors on commercial GCs.

The others are as follows:

1. Atomic-emission detector (AED)
2. Chemiluminescence detector
3. Electron-capture detector (ECD)
4. Flame-photometric detector (FPD)
5. Mass spectrometer (MS)
6. Photoionization detector (PID)

The requirements of a GC detector depend on the separation application.

For example, an analysis may require a detector selective for chlorine-containing molecules. Another analysis might require a detector that is non-destructive so that the analyte can be recovered for further spectroscopic analysis. You *cannot* use FID in that case because it destroys the sample totally. TCD, on the other hand, is non-destructive.

TCD Detector:
A TCD detector consists of an electrically heated wire. The temperature of the sensing element depends on the thermal conductivity of the gas flowing around it. Changes in thermal conductivity, such as when organic molecules displace some of the carrier gas, cause a temperature rise in the element which is sensed as a change in resistance. The *TCD is not as sensitive as other detectors but it is non-specific and non-destructive.*

ECD Detector:
Uses a radioactive beta emitter (electrons) to ionize some of the carrier gas and produces a current between a biased pair of electrodes. When an organic molecule that contains the electronegative functional group, such as halogens, phosphorus, and nitro-groups, passes by the detector, it captures some of the electrons and reduces the current.

FID Detector:
Consists of a hydrogen/air flame and a collector plate. The effluent from the GC column passes through the flame, which breaks down organic molecules and produces ions. The ions are collected on a biased electrode and produce an electrical signal. It is extremely sensitive, with large dynamic range.

MS Detector:
Uses the difference in the mass-to-charge ratio (m/e) of ionized atoms or molecules to separate them from each other. Molecules have distinctive fragmentation patterns that provide structural information to identify structural components.

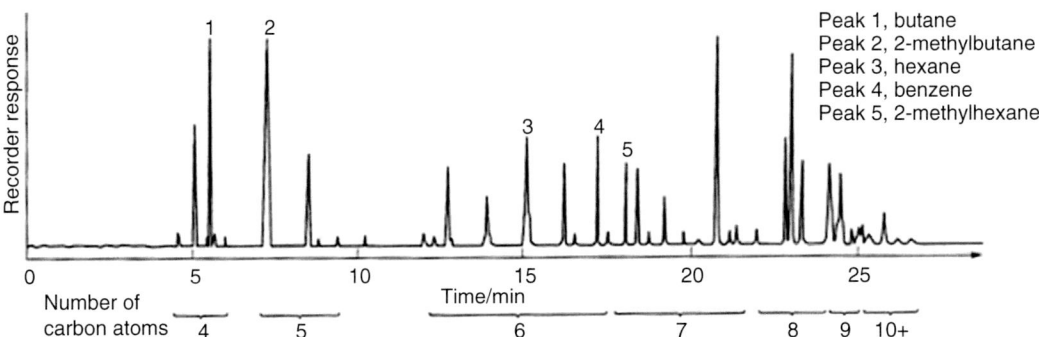

Fig. 12 Chromatogram of petrol

The general operation of a mass spectrometer is to:

1. Create gas-phase ions

2. Separate the ions in space or time based on their mass-to-charge ratio

3. Measure the number of ions of each mass-to-charge ratio

The ion separation power of an MS is described by the resolution:

$$R = m/Dm$$

Where m is the ion mass and Dm is the difference in mass between two resolvable peaks in a mass spectrum.

For example, an MS with a resolution of 1000 can resolve an ion with m/e of 100.0 from an ion with an m/e of 100.1.

Theory: *As the gas moves the solute (analyte) through and over the stationary phase, the solute will be in equilibrium with the gas and the solid phase. Since there is a mobile phase, the separation will appear as a chromatogram showing the separation of the analytes as shown in Fig. 12.*

8 Chromatogram of Petrol

Applications of GC:

Pesticides: Analysis of pesticide residues in soil, water, and food is crucial for maintaining safe levels in the environment

Food analysis: Analysis of food is concerned with the assay of lipids, proteins, carbohydrates, preservatives, flavors, colorants, and texture modifiers, as well as vitamins, steroids, drugs, pesticide residues, and trace elements.

Quality control analysis: Quality control analysis of food products can confirm the presence and quantities of the analytes, For example, fruits, fruit-derived foodstuffs, vegetables, and soft drinks, tea, and coffee were analyzed.

Food and cancer: Chemicals that can cause cancer have a wide variety of molecular structures and include hydrocarbons, amines, certain drugs, some metals, and even some substances occurring naturally in plants and molds. In this way, many nitrosamines have carcinogenic properties, and these are produced in a number of ways such as cigarette smoke. GC can be used to identify these nitro-compounds in trace quantities.

Drugs: Analytical procedures, chromatographic methods, and retention data are published for over 600 drugs, poisons, and metabolites. These data are extremely useful for forensic work and in hospital pathology laboratories to assist in the identification of drugs.

Pyrolysis gas chromatography: Pyrolysis GC (PGC) is used principally for the identification of non-volatile materials, such as plastics, natural and synthetic polymers, drugs, and some microbiological materials.

Metal chelates and inorganic materials: Organometallics other than chelates, which can be analyzed directly, include boranes, silanes, germanes, and organotin and lead compounds.

Environmental analysis: Environmental pollution is an age-old trademark of man, and in recent years as technology has progressed, populations have increased and standards of living have improved. Combustion of fossil fuel, disposal of waste materials and products, and treatment of crops with pesticides and herbicides have all contributed to the problems. Technological developments have enabled man to study these problems and realize that even trace quantities of pollutants can have detrimental effects on health and on the stability of the environment.

9 Gel Filtration Chromatography

Gel filtration (chromatography) is also known as molecular sieve chromatography. Gel filtration chromatography separates molecules according to their size and shape. The stationary phase consists of beads containing pores that span a relatively narrow size range. Smaller molecules spend more time inside the beads than larger molecules and therefore elute later (after a larger volume of mobile phase has passed through the column).

Principle:

One requirement for SEC is that the analyte does not interact with the surface of the stationary phases. Differences in elution time are based solely on the volume of the analyte. A small molecule that can penetrate every corner of the pore system of the stationary phase (where the entire pore volume and the interparticle volume ~80% of the column volume) will elute late. A very large molecule that cannot penetrate the pore system (where only the interparticle volume ~35% of the column volume) will elute earlier when this volume of mobile phase has passed through the column. The underlying principle of SEC is that particles of different sizes will elute (filter) through a stationary phase at different rates. Particles of the same size should elute together ([1]; [2]).

Procedure:

- Add the sample to the top of the resin by allowing the solution to gently run down the wall of the column.
- Place the effluent tube in the first test tube in the test tube rack (this will be fraction 1) and open the clamp.
- Do not disturb the top of the resin. Allow the sample to enter the resin and then gently add a few drops of NaCl. Allow NaCl to penetrate the column and then gently add NaCl to fill the column.
- Collect fractions until all the colored material was eluted from the column. Close the clamp. Collect 3 mL of effluent in each tube. After 3 mL has been collected in the first tube (fraction 1), switch to the second tube (fraction 2) and collect the next 3 mL, etc., as shown in Fig. 13.
- Read the absorbance at 400 nm using NaCl as blank.
- Record all your results in the table.
- Plot a graph of absorbance at 400 nm against fraction number.

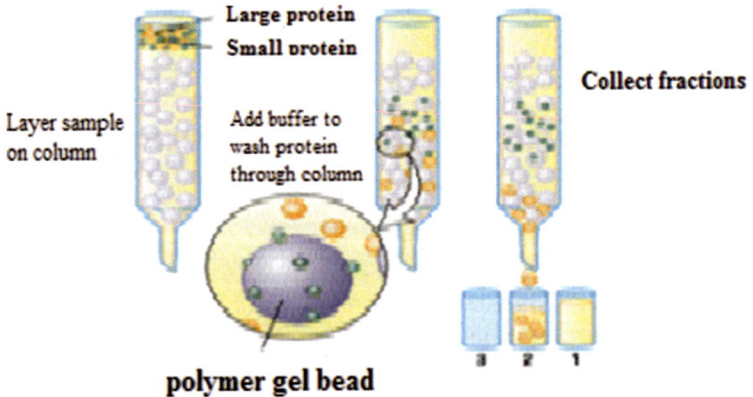

Fig. 13 Ion exchange chromatography

Types of Gels:
The gels used as molecular sieves are cross-linked polymers. They are uncharged and inert, i.e., do not bind or react with the materials being analyzed.

- Three types of gels are used:

  1. *Dextran*: It is a homopolysaccharide of glucose residues. It is prepared with various degrees of cross-linking to control pore size. It is bought as dry beads; the beads swell when water is added. The trade name is Sephadex. It is mainly used for separation of small peptides and globular proteins with small to average molecular mass.

  2. *Polyacrylamide*: These gels are prepared by cross-linking acrylamide with N,N-methylenebisacrylamide. The pore size is determined by the degree of cross-linking. The separation properties of polyacrylamide gels are mainly the same as those of dextrans. They are sold as Bio-Gel P. They are available in a wide range of pore sizes.

  3. *Agarose*: It is a linear polymer composed of D-galactose and 3,6 anhydro-1-galactose. It forms a gel that is held together by H bonds. It is dissolved in boiling water and forms a gel when it is cold. The concentration of the material in the gel determines the pore size. The pores of agarose gel are much larger than those of Sephadex or Bio-Gel P. It is useful for analysis or separation of large globular proteins or long linear molecules such as DNA.

The gel filtration material that will be used in the experiment below is called Sephadex G-75, and it will separate molecules with molecular weights from 3000 to 70,000. Molecules with molecular weights larger than 70,000 will be excluded from the beads.

The volume outside the gel matrix is known as the void volume, Vo. This is the volume required to elute a substance so large that it cannot penetrate the pores at all. Such a substance is said to be completely excluded by the gel. For Sephadex G-75, proteins with molecular weights greater than 70,000 are completely excluded. The volume of buffer required to elute any given substance is known as the elution volume, Ve, of the compound.

Applications of Gel Filtration:

- Purification of enzymes and other proteins.
- Estimation of molecular weight mainly for globular proteins: To do this, several proteins with known molecular weights are run on the column and their elution volumes determined. If the elution volumes are then plotted against the log molecular weight of the corresponding proteins, a straight line is obtained for the

separation range of the gel being used. If the elution volume of a protein of unknown molecular weight is then found, it can be compared to the calibration curve and the molecular weight determined.

Consider the separation of a mixture of glutamate dehydrogenase (MW 290,000), lactate dehydrogenase (MW 140,000), serum albumin (MW 67,000), ovalbumin (MW 43,000), and cytochrome c (MW 12,400) on a gel filtration column packed with Bio-Gel P-150 (fractionation range 15,000–150,000).

When the protein mixture is applied to the column, *glutamate dehydrogenase* would elute *first* because it is above the upper fractionation limit. Therefore it is totally excluded from the inside of the porous stationary phase and would elute with the void volume (V0). *Cytochrome c* is below the lower fractionation limit and would be completely included, eluting *last*. The other proteins would be partially included and elute in order of decreasing molecular weight.

- Determination of MW of peptides, proteins, and polysaccharides
- Desalting of colloids, e.g., desalting of albumin prepared with 2% $(NH_4)_2SO_4$
- Separation of a mixture of mono- and polysaccharides
- Separation of amino acids from peptides and proteins
- Separation of proteins of different molecular weights
- Separation of mucopolysaccharides and soluble RNA
- Separation of myoglobin and hemoglobin
- Separation of alkaloids

Advantages:

- Unlike ion exchange or affinity chromatography, molecules do not bind to the medium so buffer composition does not directly affect resolution.
- It is well suited for biomolecules that may be sensitive to changes in pH and concentration of metal ions or co-factors and harsh environments
- Conditions can be varied to suit the type of sample or the requirements for further purification, analysis, or storage without altering the separation.
- It can be used after any chromatography technology because components of any elution buffer will not affect the final separation.

Notes on the Use of Gel Filtration Chromatography:
The choice of the matrix depends on the range of size of molecules to be separated and the goal of the separation. Different bead types have pores of different sizes.

· The matrix beads normally come in dry form and must be swollen before use. It is important not to use a magnetic stirrer when preparing the beads, as it can fragment the beads. It takes several days to swell beads like Sephadex that you will use today. One shortcut, however, is to autoclave the solution. This causes the beads to swell more rapidly without damaging them.

· Never allow a gel filtration column to dry out. If it dries out, the column must be re-poured. It is crucial for good separation that the column be consistent from top to bottom (without any bubbles).

References

1. Rajan K (2011) Analytical techniques in biochemistry and molecular biology. Springer, New York. eBook ISBN978-1-4419-9785-2

2. Wilson K, Walker J (2010) Principles and techniques of biochemistry and molecular biology. Cambridge University Press, Cambridge. ISBN 978-0-521-73167-6

Chapter 6

Electrophoresis

Abstract

Many important biological molecules, such as amino acids, peptides, proteins, nucleotides, and nucleic acids, possess ionizable groups and, therefore, at any given pH, exist in solution as electrically charged species either as cations or anions. Under the influence of an electric field, these charged particles will migrate either to the cathode or to the anode, depending on the nature of their net charge. The equipment required for electrophoresis consists basically of two items, a power pack and an electrophoresis unit. This chapter deals with paper electrophoresis, SDS-PAGE electrophoresis, and immunoelectrophoresis.

Key words Paper electrophoresis, SDS-PAGE electrophoresis, Immunoelectrophoresis

Abbreviations

Ab	Antibody
Ag	Antigen
CE	Capillary electrophoresis
LVE	Low-voltage electrophoresis
SDS-PAGE	Sodium dodecyl sulfate-polyacrylamide gel electrophoresis

The differential rate of migration of ion–molecule in an electrolyte solution under the influence of an applied electric field in a support medium (e.g., paper, gel, or capillary tube). A useful method to separate substances based on their charge-to- mass ratios as shown in Fig. 1.

Mahin Basha, *Analytical Techniques in Biochemistry*, Springer Protocols Handbooks,
https://doi.org/10.1007/978-1-0716-0134-1_6, © Springer Science+Business Media, LLC, part of Springer Nature 2020

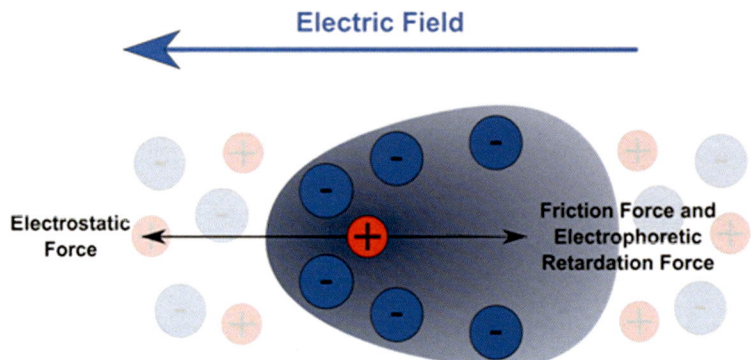

Fig. 1 Motion of a charged particle by electrophoresis

Principle:

Charged ion or molecule migrates when placed in an electric field

Rate of migration depends on its net charge, size, shape and the applied electric current

$$v = \mu_e E$$

where, v = velocity of an ion

E = electric field strength (Vcm^{-1})

μ_e = electrophoretic mobility

= distance migrated in a certain time period

The electrophoretic mobility is given by

$$\mu_e = \frac{q}{6\pi\eta r} \text{ (when electric force = frictional drag)}$$

showing that small highly charged species have high mobility and vice versa.

Driving Force of Migration: *Resultant of the electrostatic force of attraction between the electric field and the charged molecule and the retarding forces due to friction and electrostatic repulsion from molecules of the transport medium as shown in Fig. 2.*

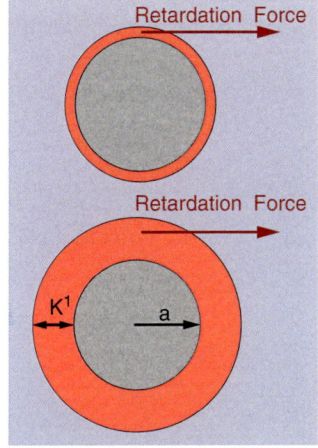

Fig. 2 Illustration of electrophoresis retardation

---

1   Supporting Media for Electrophoresis

Paper:

– Filter paper such as Whatman no. 1 and no. 3 MM
– Used to good effect

Cellulose Acetate:

– Containing two to three acetyl groups
– To give sharper bands
– More easily rendered transparent
  Low solvent capacity
  Enhancing the resolution

Gels:

– Three-dimensional semisolid colloids
– Resolving power enhanced due to sieve effect operating
– Prepared from starch, agar, or polyacrylamide

Procedure:

• Immersion of two electrodes in two separate buffer chambers but not fully isolated from each other
• Migration of charged particles from one chamber to the other by using an electric field
• Separation of different ions migrating at different speeds

2   Factors Affecting Electrophoretic Mobility

Charge – The higher the charge, the greater the mobility.

Size – The bigger the molecule, the greater the frictional and electrostatic forces exerted on it by the medium, i.e., larger particles have smaller electrophoretic mobility compared to smaller particles.

Electric field – An increase of migration with the increase of voltage gradient.

Buffer – Dependence of migration on pH of the buffer.

Ionic strength – The greater the ionic strength of the buffer solution, the higher the proportion of the current and hence electrophoretic mobility.

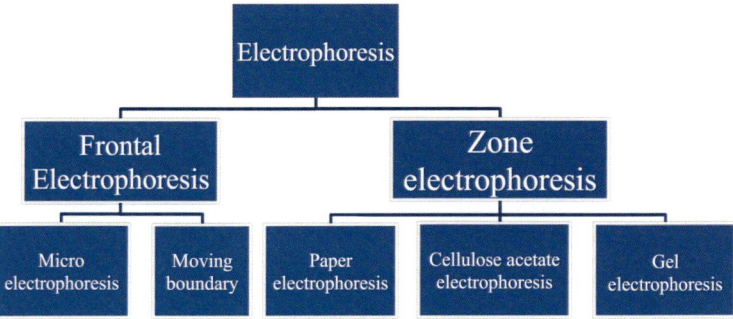

3   Structure Shows Types of Electrophoresis

Techniques of Electrophoresis:

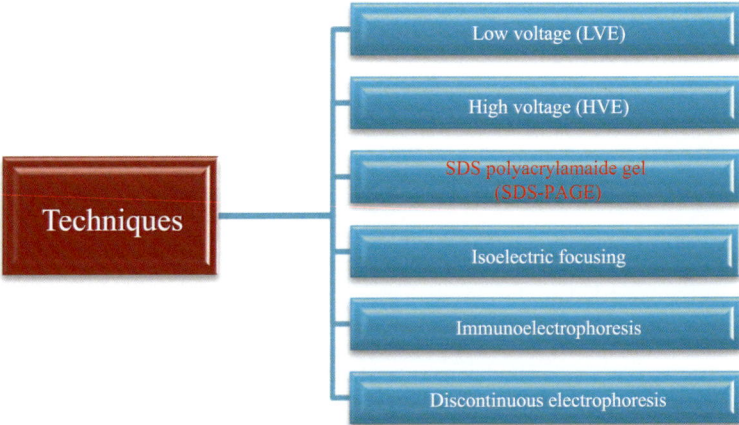

Applications of Electrophoresis:

- DNA analysis
- Protein analysis
- Antibiotic analysis
- Vaccine analysis
- Detection of damaged genes by gel electrophoresis
- To be used in forensic research

Low-Voltage Electrophoresis: Two compartments to hold the buffer and electrodes. A suitable carrier for support medium ending in contact with the buffer medium. To provide voltage gradient ~5 Vcm^{-1}, a power pack supplying up to 500 V or even 1000 V and 0–150 mA.

Application of LVE:

- To separate any ionic substances
- The examination of biological and clinical specimens for amino acids and proteins as shown in Fig. 3
- Separation of sugars

High-Voltage Electrophoresis:

- To obtain voltage gradients up to 100 Vcm^{-1}, high voltage and current supplying
- Using cooling plates for heat dissipation generated by high voltage
- Less than of 1 h analysis time
- Working best with small ions deriving from small peptides and amino acids

Capillary Electrophoresis (CE): Separation of analyte species achieved on the basis of differential migration in an electric field through narrow-bore fused-silica capillary columns (25–100 μm).

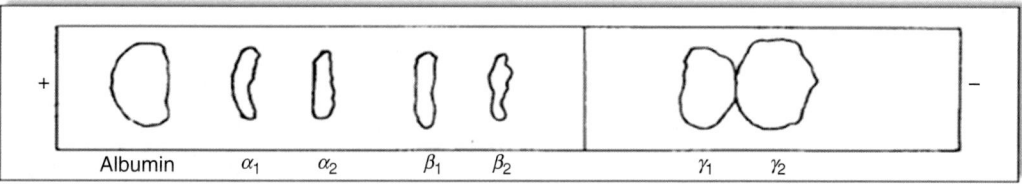

Fig. 3 Electrophoretogram of plasma proteins on cellulose acetate at pH 8.6

Overview of the Instrumentation of CE: A fused capillary column dipping into two electrolyte buffers containing Pt foil cathode or anode across 15–60 kV voltage applied. Introducing a small volume of sample at one end of the capillary. Migration of sample through the capillary under the force of applied electric field.

Advantages of CE:

- Power dissipation minimized by high electrical resistance
- Having voltage gradients up to 100–500 Vcm^{-1} necessary for rapid separations
- Most prominently used because of its faster results and high-resolution separation
- Large range of detection methods available
- No Joule–Thomson effect
- No band broadening

4 Paper Electrophoresis

Paper electrophoresis is one of the zone electrophoreses. This is a very important method in all laboratories. In this article let us learn the details of paper chromatography with suitable notes. I have given the info about this in Notes [1, 2].

Principle: "The charge carried by a molecule depends on the pH of the medium. Electrophoresis at low voltage is not usually to separate low molecular weight compounds because of diffusion, but it is easier to illustrate the relationship between charge and pH with amino acids than with proteins (or) other macromolecules."

Filter Paper: Paper of good quality should contain at least 95% α-cellulose and should have only a very slight adsorption.

Apparatus: The equipment required for electrophoresis consists basically of two items, a power pack and an electrophoretic cell.

Power pack provides a stabilized direct current and has controls for both voltage and current output, which have an output of 0–500 V and 0–150 mA.

The electrophoretic cell contains the electrodes, buffer reservoirs, a support for paper, and a supporting transparent insulating cover. The electrodes are usually made of platinum.

Two arrangements of the filter strips are commonly used: the horizontal and vertical arrangements. Both the arrangements are equally viable and the choice usually depends upon personal preferences as shown in Fig. 4.

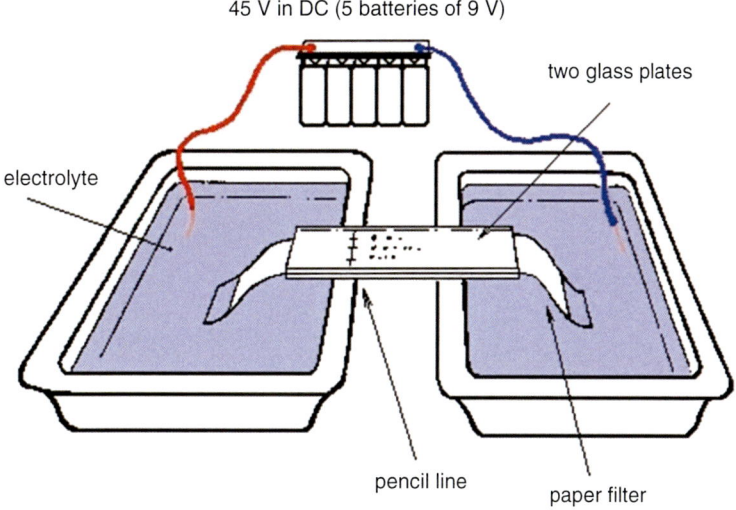

Fig. 4 Apparatus for paper electrophoresis

Table 1
Commonly used buffers for electrophoresis

Intended separation	Buffer	pH	Ionic strength	Composition liter
Proteins	Barbital	8.6	0.05	10.3 g Sodium barbiturate 1.84 g. Barbital
	Phosphate	7.4	...	0.6 g. $NaH_2PO_4.H_2O$ 2.2 g. Na_2HPO_4
Nucleoproteins	Acetate	4.5	0.1	3.51 g. NaCl 3.28 g.Sodium acetate (pH adjusted with HCl to 4.5)
	Citrate	4.5	0.13	28.46 g Sodium citrate 20.6 g. Citric acid
Amino acids	Phosphate	4.6	0.15	20.4 g KH_2PO_4
	Michalies	8.6	0.1	9.8 g Sodium barbiturate

Sample Application: The sample may be applied as a spot (about 0.5 cm in diameter) or as a narrow uniform streak.

Special devices are available commercially for this purpose. They can be applied before the paper has been equilibrated with buffer or after it.

Procedure: After the sample has been applied to the paper and the paper has been equilibrated with the buffer, the current is switched on. Commonly used buffers are barbital, phosphate, acetate, and citrate as shown in Table 1.

The device providing stable voltage or current is available. A frequent observation is necessary to run an electrophoretic apparatus. Overheating can be avoided by placing the entire equipment in the cold room. The process does not take longer than 2 h. After 2 h the power is switched off and the paper is removed. Once removed, the paper is dried in a hot oven at 1100 °C.

5 Detection and Quantitative Assay

To identify the unknown electrophoretogram, compare the unknown electrogram with standard electrogram under standard conditions. Individual compounds are usually identified by physical properties by the following methods:

(i) *Fluorescence:*

    (a) Staining with "ethidium bromide" and subsequent visualization of the electrophoretogram under UV light makes DNA and RNA fluoresce and thus facilitates their detection.

    (b) Fluorescamine staining is utilized for detecting amino acids, amino acid derivatives, peptides, and proteins.

    (c) Dansyl chloride may be used in place of fluorescamine.

(ii) *UV absorption:*

Proteins, peptides, and nucleic acids absorb in the range of 260–280 nm; this property can be detected.

(iii) *Staining:* Dye used in chromatography is shown in Table 2.

(iv) *Detection of enzymes in situ:*

If the component to be separated is an enzyme, special techniques may be used to detect it.

The paper strip, which has separated enzyme, is impregnated with the substrate for the enzyme desired to be separated.

Table 2
Staining in chromatography

Compound	Dye	Comments
Proteins	Bromophenol blue in acetic acid DANSYl chloride Fluorescamine	Visual, quantitative Fluorescent, Quantitative Fluorescent, Very sensitive
Nucleic acids	Methyl green-pyronine Ethedium bromide	DNA-Blue, RNA-Red, Sensitive Fluorescent, Very sensitive
Polysaccharide	Iodine	Visual, Sensitive
Lipoprotein	Sudan black in 60% ethanol	Visual, Sensitive
Glycoprotein	Alcian blue	Visual, Sensitive

The paper strip is now placed in a suitable buffer along with electrophoretogram. The color bands will appear which indicates the position of the enzyme.

(v) *Quantitative estimation:*

The color density of the zone may be multiplied with the area of the zone, and the resulting value would be a rough estimate of the concentration of the component.

Applications:

- Serum analysis for the diagnostic purpose is routinely carried about by paper electrophoresis.

- Analysis of muscle proteins, egg white proteins, milk proteins, and snake and insect venom is done by this technique.

6 SDS-Polyacrylamide Gel Electrophoresis

Sodium dodecyl sulfate-polyacrylamide gel electrophoresis (SDS-PAGE) is a technique widely used in biochemistry, forensics, genetics, and molecular biology to separate and identify proteins *according to their molecular weight*. This method separates proteins based primarily on their molecular weights [1, 2].

Principle:

- Sodium dodecyl sulfate [SDS]: is a detergent which denatures proteins by binding to the hydrophobic regions; all non-covalent bonds will be disrupted and the proteins acquire a negative net charge.

- A concurrent treatment with a disulfide reducing agent such as β-mercaptoethanol or DTT (dithiothreitol), which further denatures the proteins by reducing disulfide linkages, thus overcoming some forms of tertiary protein folding and breaking up quaternary protein structure.

- So, the protein samples are having uniformed structure and charge → the separation will depend on their molecular weight only.

- Small proteins migrate faster through the gel under the influence of the applied electric field.

- The number of SDS molecules that bind is proportional to the size of the protein, Thereby in the electrical field, protein molecules move toward the anode (+) and are separated only according to their molecular weight.

- The protein samples are having uniformed structure and charge → the separation will depend on their molecular weight only.

- SDS-treated proteins have very similar charge-to-mass ratios and similar shapes. During PAGE, the rate of migration of SDS-treated proteins is effectively determined by molecular weight.

- Small proteins migrate faster through the gel under the influence of the applied electric field, whereas large proteins are successively retarded, due to the sieving effect of the gels as shown in Figs. 5, 6 and 7.

Polyacrylamide Gel: The polyacrylamide gel is formed by copolymerization of acrylamide and a cross-linking

By N,N'-methylenebisacrylamide or "bisacrylamide." To polymerize the gel, a system consisting of ammonium persulfate

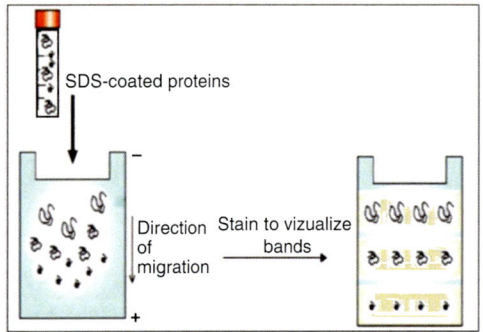
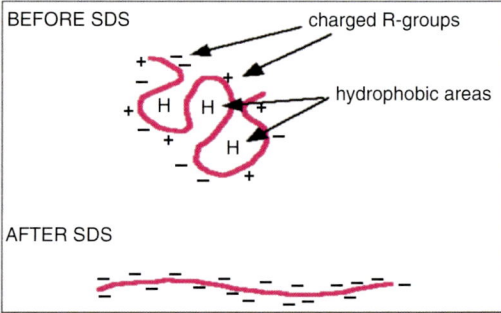

Fig. 5 SDS-polyacrylamide gel electrophoresis

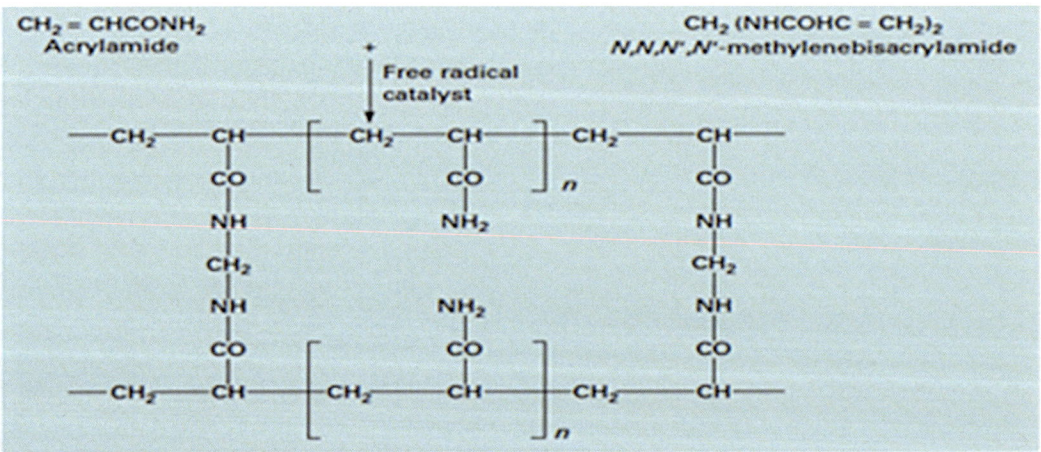

Fig. 6 The formation of polyacrylamide gel from acrylamide and bis-acrylamide

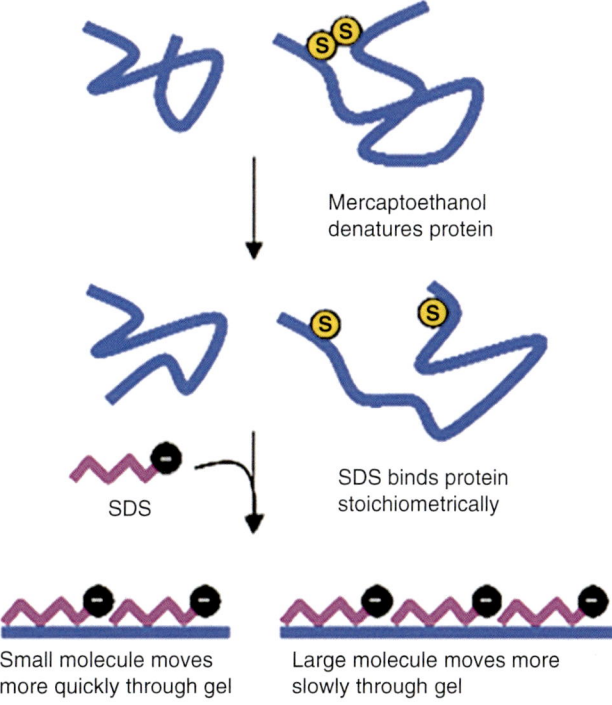

Fig. 7 Schematic diagram of SDS-PAGE separation of proteins

(initiator) and tetramethylene ethylenediamine (TEMED) is added [catalyst].

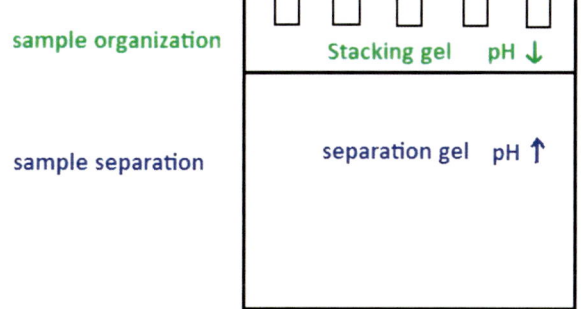

7   SDS-Polyacrylamide Gel Electrophoresis

SDS-Polyacrylamide Gel Electrophoresis Preparations:

1. *Sample preparation:* 40 µl of protein sample + 10 µl of *disruption buffer* → *boil* the mixture for 3 min at 99 °C
     SDS-PAGE – Disruption buffer contains:
   – 10% (w/v) SDS
   – 1M Tris/HCl, pH 6.8

- Glycerol
- β-Mercaptoethanol
- Bromophenol blue

2. *Polyacrylamide gel preparation:*
    Acrylamide stock should be prepared first:

- Cross-linked polyacrylamide gels are formed from the polymerization of acrylamide monomer in the presence of smaller amounts of N,N'-methylenebisacrylamide (normally referred to as "bisacrylamide").

(A) *Separation gel preparation*

Stock solutions	Volume of stock solution required to make 12% polyacrylamide gel
1.5 M Tris/HCl, pH 8.8	2.0 ml
Acrylamide stock	3.2 ml
Water	2.8 ml
10% SDS	80 µl
10% Ammonium persulfate (fresh)	100 µl
TEMED	20 µl

(B) *Stacking gel preparation*

Stock solutions	Volume of stock solution required to make 12% polyacrylamide gel
0.5M Tris/HCl, pH 6.8	1.0 ml
Acrylamide stock	1.0 ml
Water	3.0 ml
10% SDS	80 µl
10% Ammonium persulfate (fresh)	100 µl
TEMED	20 µl

3. *Run the gel using a running buffer 1x pH 8.4:*
    It contains:

- Tris-HCl
- Glycine
- SDS

4. *Stain the gel using a staining buffer:*

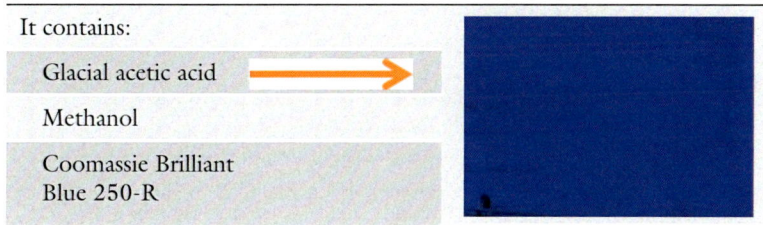

It contains:	
Glacial acetic acid	⟶
Methanol	
Coomassie Brilliant Blue 250-R	

5. *De-stain the gel using a de-staining buffer:*

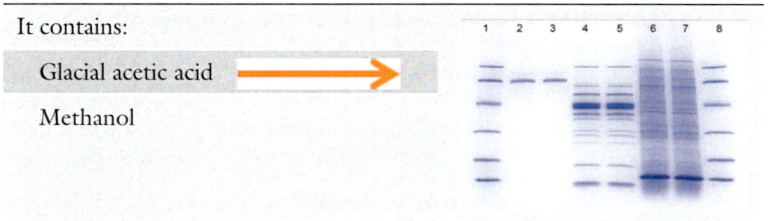

It contains:	
Glacial acetic acid	⟶
Methanol	

Applications:

1. To detect the purity of the protein
2. To determine protein molecular weight

8   Immunoelectrophoresis

The principle of immunoelectrophoresis is the formation of precipitate lines (insoluble immunocomplexes) at the point of equivalence (zone of equivalence) of the antigen and its corresponding antibody. This technique combines the specificity of immunoprecipitation reactions with the separation of molecules by electrophoresis in a molecular sieving medium. Usually, the analysis is carried out in an agarose gel containing barbitone buffer on a microscopic slide. In this technique, it is important that the ratio between the quantities of Ag and Ab be correct (antibody titer). When the amount of antibody is in excess, statistically at most one Ag molecule binds to each molecule of Ab. When the amount of Ag is in excess, at most one molecule of Ab binds to each Ag molecule. However, at a specific Ag/Ab ratio (equivalence point), an optimal amount of Ag–Ab immunocomplexes (macromolecules) is formed. They consist of an Ag–Ab–Ag–Ab network and are immobilized in the gel matrix because of their bulky size as an immunoprecipitate. The white/milky precipitate lines are visible in the gel and can be revealed with protein stains. The agarose gels are used in

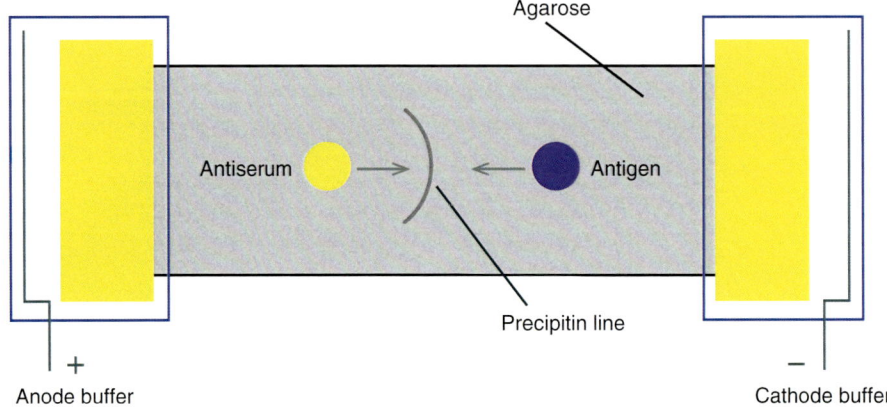

Fig. 8 The schematic diagram of counterimmunoelectrophoresis

immunoelectrophoresis and the separation times are exceedingly low: about 30 minutes. The method is highly specific and the sensitivity is also very high because distinct zones are formed. Immunoelectrophoresis can be categorized into three principles [1, 2].

8.1 Counterimmuno-electrophoresis

In an agarose gel exhibiting high electro-osmosis, the buffer is set at a pH of about 8.6 so that Ab does not carry any charge. A suitable pattern is cut with a gel punch and 1–10 microliters of solutions containing 1–100 micrograms of antigen are added to the wells. Thick wet filter paper wicks to the electrode buffer connect the slides, and a direct electric current of about 8 mA per slide is passed for 1–2 h. The sample and the Ab placed in opposite wells move toward each other, the negatively charged Ags migrate electrophoretically, and the antibodies are carried by electro-osmotic flow (Fig. 8). The interaction of Ag with the Ab results in the formation of precipitation lines. This technique is more rapid (15–25 min) than the Ouchterlony method, which may take days to produce a clear result. Also, this technique is more sensitive because all of the molecules migrate toward each other rather than diffusing radially. This method has been used and described by Estela and Heinrichs (1978). The technique is especially useful in forensic science for establishing the origin of body fluids such as blood, semen, and saliva.

8.2 Zone Immunoelec-trophoresis

According to Grabar and Williams (1953) firstly a zone electrophoresis is run in an agarose gel, followed by the diffusion of the Ag fraction toward the Ab which is pipetted into rectangular narrow troughs cut in the side parallel to the electrophoretic run. The

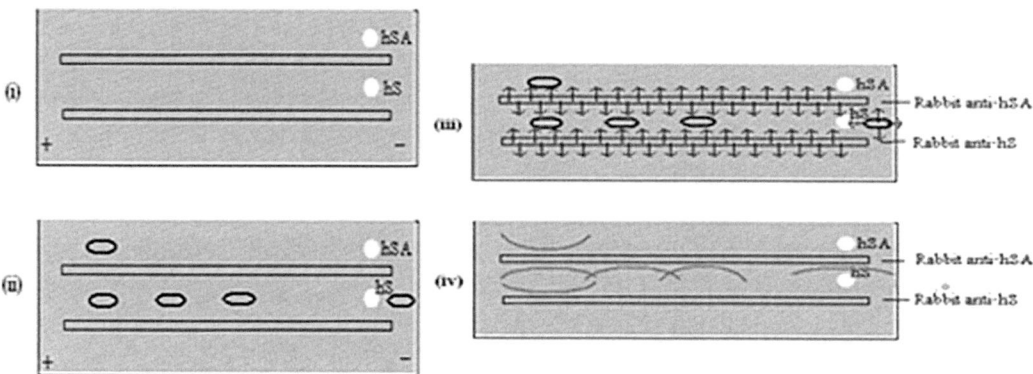

Fig. 9 Various steps in zone immunoelectrophoresis. (i) Round wells punched in the agarose are filled with Human Serum Albumin (HSA) and Human Serum (HS); (ii) After electrophoresis the proteins will migrate as depicted but will not be visible. Rectangular troughs are cut in the agarose and agarose is removed from troughs; (iii) Troughs are filled with the antisera, the antigens and antibodies diffuse through the porous agarose as indicated by arrows; (iv) Finally, the precipitin lines appear

charged molecules will be separated electrophoretically but they will not be visible (Fig. 9). Immediately after the voltage supply has been disconnected, the troughs are filled with an appropriate antiserum and incubated overnight at room temperature in a humid chamber. The antigens diffuse radially and the antibodies diffuse laterally resulting in the Ag–Ab precipitation arcs. Despite the use of agarose, which acquires a smaller charge than agar, the electro-osmotic flow of water during electrophoresis moves all of the antigens toward the cathode. This results in an apparent cathodic migration of gamma globulins, including IgG antibodies. This technique may be used to determine the purity of or detect a particular antigen in sera, culture filtrates, tissue or cell extracts, or fractions from any preparative procedure.

8.3 Laurell's Rocket Electrophoresis

In Laurell's rocket electrophoresis (Laurell 1966) and related methods, Ags migrate in an agarose gel, which contains a definite concentration of Ab. As in the above method, the Abs are not charged because of the choice of the buffer. As the sample migrates one Ab will bind to one Ag until the ratio of concentrations corresponds to the equivalence point of the immunocomplex. The result is that the rocket-shaped precipitation lines are formed (Fig. 10); the enclosed areas are proportional to the concentration of Ag in the sample. A series of modifications to this technique exist, including two-dimensional ones.

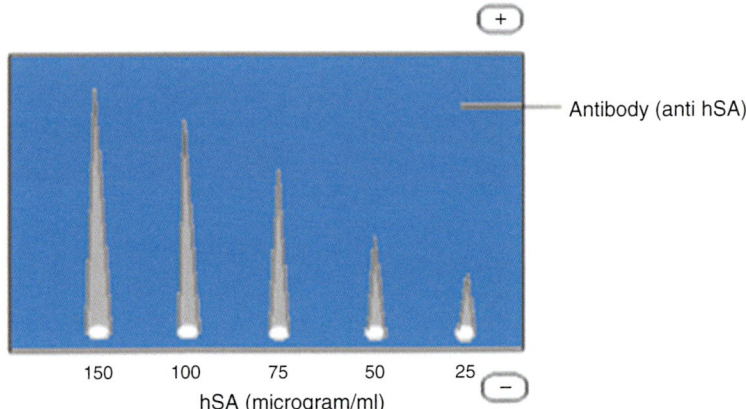

Fig. 10 A schematic diagram of Laurell's rocket immunoelectrophoresis. The agarose gel contains anti-human Serum Albumin (HSA) antibody and the wells punched in the agarose contain varying concentrations of the Human Serum Albumin (HSA). The electric fiels is applied to obtain the rocket like precipitin bands as shown above

References

1. Rajan K (2011) Analytical techniques in biochemistry and molecular biology. Springer, New York. eBook ISBN978-1-4419-9785-2

2. Wilson K, Walker J (2010) Principles and techniques of biochemistry and molecular biology. Cambridge University Press, Cambridge. ISBN 978-0-521-73167-6

Chapter 7

Immuno-techniques

Abstract

The vast majority of immunoassays carried out fall into the category of enzyme immunosorbent assays. These are routinely used for the diagnosis of infectious agents such as viruses and other substances in blood. The antigen is the substance or agent to be measured. In this technique the antigen is immobilized onto a solid phase, either a reaction vessel or a bead.

An antigen is a substance capable of causing an immune response leading to the production of antibodies, and it is also a target to which antibodies will bind. Antibodies are antigen specific and will only bind to the antigen that initiated their production. The purpose of this chapter is to discuss antibody and antigen reactions that will be focused on various immuno-techniques.

Key words Immuno-techniques, Immunodiffusion, Immuno-fluorescence, Enzyme-linked immunosorbent assay (ELISA), Radioimmunoassay (RIA), Western blotting or immunoblotting

Abbreviations

BSA	Bovine serum albumin
FACS	Fluorescence-activated cell sorter analysis
FCA	Freund's complete adjuvant
FIA	Freund's incomplete adjuvant
ISCOMs	Immune-stimulating complexes
SPIEM	Solid-phase immune electron microscopy
SRID	Single radial immunodiffusion

1 Introduction

The basis of antigen–antibody reaction has been extensively exploited to devise techniques to qualitatively and quantitatively detect the presence of different biomolecules with high sensitivity and specificity. Any biomolecule that has a complex structure, is protein in nature, and possesses high molecular mass (> 1000 daltons) and internal complexity when injected into an animal body produces specific proteins termed as antibodies. The complex carbohydrates (bacterial LPS or lipopolysaccharides) and nucleic acids

Mahin Basha, *Analytical Techniques in Biochemistry*, Springer Protocols Handbooks, https://doi.org/10.1007/978-1-0716-0134-1_7, © Springer Science+Business Media, LLC, part of Springer Nature 2020

(RNA and DNA) are also immunogenic albeit to a lesser extent. The antibodies have the inherent property to bind to the antigen used to elicit antibody response with high avidity and sensitivity. The antigens encompass the biomolecules that are part of bacteria, fungi, protozoan parasites, etc. The antigens also include smaller molecules (< 1000 daltons) such as p-nitrophenol and other purely chemical molecules that are known as haptens. Haptens are made immunogenic by their binding with carrier molecules such as bovine serum albumin (BSA) and keyhole limpet hemocyanin (KLH) molecules to produce modified antigens. Thus theoretically the injection of an exogenous protein that is foreign to an animal (host) into its tissue through a suitable route (intramuscularly, subcutaneous, and intraperitoneal) elicits formation of specific antibodies. The process of injection of a foreign antigen into an animal host is called immunization. The immunization involves a series of injections of a given antigen by a suitable route periodically over a period of time. The method of immunization that generates a detectable and a sustainable immune response in the host is called primary immunization. When an antigen is injected into the host, the IgM class of antibodies are the first to develop, followed by the gradual development of IgG antibodies. During a natural infection, the detection of an antigen-specific IgM antibody indicates a recent history of infection. However, if antigen-specific IgG antibody preexisted in the host, a quite early history of a natural infection or immunization is often seen. The ability of the antigen to produce an enhanced and prolonged immune response by formation of specific antibodies (humoral immunity) is achieved by combining/mixing a given antigen with non-specific immuno-stimulatory compounds called adjuvants. The judicious use of an adjuvant is essential to induce a strong antibody response to soluble antigens. The adjuvants include aluminum phosphate, aluminum oxide, and Freund's complete (FCA) and Freund's incomplete (FIA) adjuvants. For developing a state of immunity in human and animals, the antigens are mixed with aluminum phosphate and aluminum hydroxide (alum), and in experimental animals, FIA or FCA might be used. Freund's adjuvant should be used when a small amount of the immunogen is available for immunization. If a large amount is available or if the compound is known to be highly immunogenic, then other adjuvants can be used. However, FIA or FCA must never be given intravenously. The use of ISCOMs (immune-stimulating complexes) and Quil-A (sourced from *Quillaja* plant) as adjuvants has also been reported. The purpose of any immunization protocol using an antigen is to achieve a strong protective immunity. The animals such as guinea pig, rabbit, horse, mouse, and rats are often used to generate specific polyclonal antibodies. However, the use of hybridoma technique developed. It can produce the bulk amount of specific monoclonal antibodies in vitro [1, 2].

The antigen-specific IgM or IgG class of antibodies can be purified in large amounts by use of affinity chromatography. Protein A, a protein of *Staphylococcus aureus* cell wall, has a natural inherent ability to bind to IgG antibody of a variety of animals and humans. Protamine has a natural affinity to bind IgM antibody. The ligands like protein A or protamine when covalently (stable) attached to support matrices (dextran such as Sephadex, Sephacryl, and Sepharose and cellulose) provide an efficient tool to purify large amounts of IgG or IgM antibodies for analytical purposes. Furthermore, use of matrix-bound antigen provides easy affinity separation/purification of antigen-specific antibodies from serum obtained from a previously immunized animal or ascites obtained from mice injected with hybridoma producing specific antibodies. These purified monoclonal antibodies or polyclonal antibodies can be generated for use in highly specific and sensitive techniques like immunodiffusion, immuno-electrophoresis, ELISA, Western blotting, immuno-fluorescence microscopy, fluorescence-activated cell sorter (FACS) analysis, solid-phase immune electron microscopy (SPIEM), and many clinical diagnostic tests for detection of bacterial (Widal test for detection of *Salmonella typhi*/typhoid), mycoplasma, and viral infections (hepatitis A and hepatitis B viruses). On the other hand, the presence of antigen-specific IgM or IgG antibody in the body fluid/blood suggests a recent or a previous exposure to a pathogenic bacterial or viral agent (hepatitis C, hepatitis E, and HIV). The availability of highly specific antibodies against a range of antigens is a big advantage to detect an array of pathogenic microbes as well as cancers and tumors.

The number of electrophoretic separation methods has increased dramatically since Tiselius' pioneer work for which he received the Nobel Prize in 1948. Development of these methods has progressed from paper, cellulose acetate membrane, and starch gel electrophoresis to molecular sieving, disc-PAGE, SDS-PAGE, and immuno-electrophoresis and finally to Western blotting, Southern blotting, ELISA, and FACS analysis [1, 2].

2 Immunodiffusion

For easy diffusion of relatively bigger molecules such as antigens (Ag) and antibodies (Ab; size > 150 kDa), the frictional resistance of the gel is kept negligible so that the movement of the biomolecules is not hindered and these molecules can easily diffuse toward each other when separated in space. Agarose gels with a concentration of 0.7–1.0% are often used in clinical laboratories for the analysis of antigens and antibodies and also for their quantitative analysis. This technique is performed at neutral pH; antigen and antibody solutions are poured into wells created in the solidified agarose on a glass slide. The wells of uniform diameter and depth are cut in the gel and filled individually with an antigen or an antibody solution. When molecules such as soluble Ag diffuse from a homogenous solution into an agarose gel, the concentration falls from a maximum at the solution/gel interface to zero at the leading edge of the region penetrated. Thus the system rapidly adjusts to provide a complete antigen concentration gradient. Somewhere along this concentration gradient will be an antigen concentration that will give equivalence with any given concentration of Ab. Based on this concept, a range of highly sophisticated immunodiffusion assays have been developed to detect and quantitate Ab and Ag.

Ag and Ab diffuse toward each other through the porous agarose gel, and on physical contact Ag and Ab if they recognize each other (are specific) tend to form insoluble immune complexes (precipitates) that appear as a milky band (precipitin line/band) that can be seen with the naked eyes. The precipitin band appears at the point of equivalence where the concentration of antigen and the specific antibody is equivalent. The non-precipitated proteins can be washed out. The precipitin band can be made prominent by fixing and staining (using Amido Black or Coomassie Brilliant Blue R-250 dye). The method requires 24 h or more for detection of a specific antigen and antibody reaction [1, 2].

2.1 Single Immunodiffusion

This method involves the diffusion of Ag from a solution into an agarose gel containing Ab. Mancini first described this technique known as single radial immunodiffusion (SRID). In this technique, a range of antigen concentrations is poured in wells cut in agarose gel containing corresponding antibody solution on a glass slide. The Ag diffuses radially through the porous gel out of the well resulting in the formation of a precipitin line that appears to move outward, eventually becoming stationary at the zone of equivalence where Ag complexes optimally with the Ab. The precipitin band diameter at equivalence is a function of the antigen concentration. By plotting precipitin ring diameter or circular area at equivalence against antigen concentration, a reference curve may be plotted for determining the concentration of antigen in unknown solutions. This technique is commonly used for determining the concentrations of various plasma proteins such as IgG and IgM in patients suspected to be suffering from a gamma-globulinemia and multiple myeloma, respectively, as shown in Fig. 1.

2.2 Double Immunodiffusion in Two Dimensions (Ouchterlony Technique)

In this technique, both Ag and Ab diffuse toward each other through the porous agarose. Each of the reactants develops a concentration gradient having the highest concentration close to the periphery of the well. A 2-mm-thick layer of the agarose gel is prepared on a rectangular glass slide. The circular wells (which are 10–15 mm apart) are punched in the agarose gel with a gel punch.

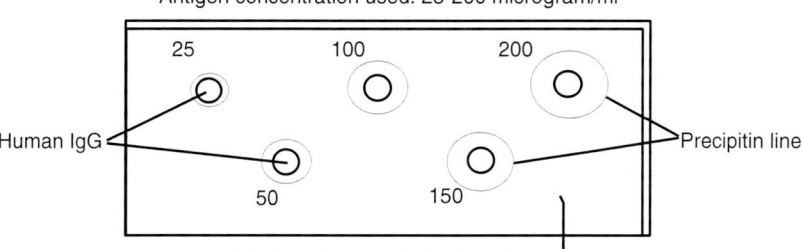

Fig. 1 Single radial immunodiffusion for estimation of human IgG

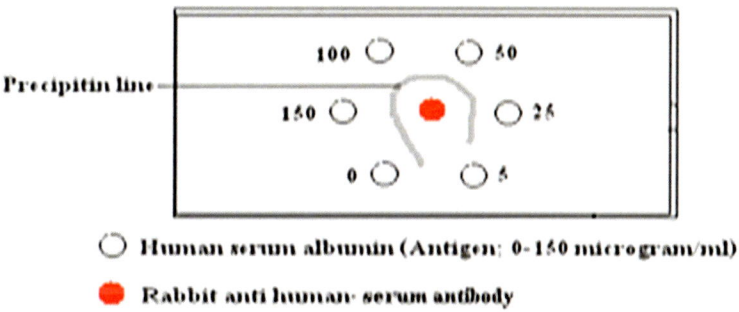

Fig. 2 Schematic diagram of double immunodiffusion in two dimensions using varying concentrations of antigen (human serum albumin) and a specific rabbit antibody (anti-human-serum antibody)

A typical pattern used to compare different antigen fractions is shown in Fig. 2. The peripheral wells are filled with different concentrations of the same Ag (0–150 microgram/ml of human serum albumin), and the Ab (rabbit anti-human-albumin antibody) is added in the central well. This slide containing the Ag and Ab solutions is kept in a humid chamber at 8 °C for 24–48 h. The pattern of the precipitin line so formed is observed. No precipitin band is formed toward the well-containing buffer alone (without antigen).

The antiserum used in the center wells contained antibodies to most of the components of the human serum; the presence of a single precipitin line between this well and the peripheral wells indicates that these wells contained an apparently pure human albumin that does not appear to be contaminated with any other human serum protein. It shall, however, be noted that this test does not provide information about any contaminant to which the center well contains any antibodies as shown in Fig. 2.

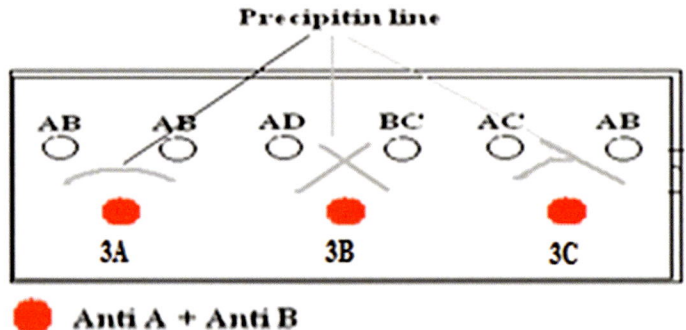

Fig. 3 (**a**, **b**, **c**) Various patterns of precipitation (Ouchterlony technique) formed during double diffusion in two dimensions. (**a**) Reaction of identity (see two lines are fused); (**b**) Reaction of non-identity (see two precipitin lines cross); (**c**) Reaction of partial identity (see right line spurs). A, B, C, and D represent antigenic determinants. Antiserum A + Antiserum B was used in each of the depicted reactions

A reaction of identity occurs between an Ab and Ag containing identical antigenic determinants and produces smoothly fused precipitin band (Fig. 3a). A reaction of non-identity occurs when the antiserum contains antibodies to both antigens but the two antigens do not share a common determinant. The two precipitin lines are formed independently with different antibody molecules and cross without interaction (Fig. 3b). A reaction of partial identity occurs when two antigens have at least one common determinant, but where the antiserum contains antibodies to a determinant in one antigen that is absent from the other (Fig. 3c).

The relative position of a precipitin line provides an estimate of antigen concentration, i.e., the more concentrated the antigen preparation, the more the distance at which the precipitin line will be formed (Fig. 3a–c). The shape of a precipitin line gives a rough estimate of the relative molecular mass of a globular protein antigen. The major antibody is usually IgG with a relative molecular mass of 15 kDa. Globular proteins with relative molecular masses significantly less than 150 kDa will diffuse through the gel more rapidly and will yield a precipitin band facing the concavity toward the low molecular mass reactant (antigen). Those with molecular weights significantly greater than 150 kDa will diffuse more slowly and produce a curved precipitin line in the opposite direction. Antigens with relative molecular masses similar to the IgG shall produce a straight precipitin band.

Thus double immunodiffusion technique is used for detection of antibody-specific antigens in the immuno-sera, chromatographic fractions, and cell fractions and to obtain information whether two antigens are identical or different or may share common antigenic determinants as shown in Fig. 4a–c.

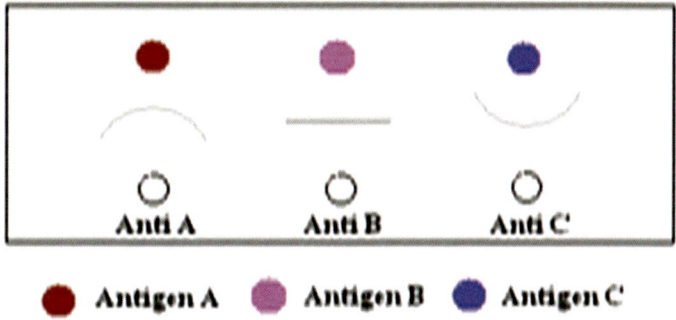

Fig. 4 (**a**, **b**, **c**) The effect of the relative molecular size of an antigen on the shape of the precipitin lines formed in double immunodiffusion (Ouchterlony technique)

3 Immuno-fluorescence

The antigens and antibodies that possess an inherent property of specificity for each other when attached/tagged with a suitable fluorescence molecule could be used to detect or localize the cells bearing surface receptor of interest or detect specific cell type. This type of fluorescence is called extrinsic fluorescence which can be detected with a fluorescence measuring device (fluorimeter or a UV/fluorescence microscope). Sometimes the fluorescence molecule (propidium iodide binds to DNA in dead cells; Evans Blue binds to non-fluorescence cells) may be able to bind intracellular organelle or some cell component without the use of an antibody – this is referred to as intrinsic fluorescence – and this stain is used as a counterstain. The fluorescence molecules include fluorescein, rhodamine, and Texas Red. The properties of these fluorochromes are summarized in Table 1. These molecules when excited with suitable wavelength absorb energy and get excited. The absorbance of energy by the fluorescence moiety occurs in less than 10^{-15} s. The excited molecule then falls back to the ground state with the release of energy in the relatively longer wavelength range in a very short time (less than 10^{-8} s). The released energy is detected with the help of detectors. The fluorescence can be visually seen through a UV/fluorescence microscope. The cells bound to the antibody tagged with a fluorescence molecule appear to be distinctly colored. The same cell may be visualized to highlight different surface receptors, intracellular organelles, etc. by using more than one fluorescence-tagged antibody. However, such detection is possible when the fluorescence molecules attached to different antibodies (with different specificities) are capable of excitation at the same wavelength but emit rays at different wavelengths. The cells that have been fixed on the microscope slide or suspended (live) cells or a thin layer of tissue(s) can be subjected to immuno-fluorescence

Table 1
Properties of fluorochromes used in cell staining

Fluorochrome	Excitation wavelength (nm)	Emission wavelength (nm)	Color
DAPI	365	>420	Blue
Fluorescein	495	525	Green
Hoechst 33258	360	470	Blue
R-phycocyanin	555, 618	634	Red
B-phycoerythrin	545, 565	575	Orange, red
R-phycoerythrin	480, 545, 565	578	Orange, red
Rhodamine	552	570	Red
Texas Red	596	620	Red

staining using appropriate antibodies attached to fluorescence molecules. The fluorescent cells are seen in the dark to maintain high efficiency of detection as well as to minimize the effect of stray radiations that might influence the results.

$$E_{excitation} - E_{emmision} = \Delta E$$

$$\text{Or } h\nu_{excitation} - h\nu_{emission} = \Delta h\nu$$

$$\text{Thus } E_{emmision} < E_{excitation}$$

The energy associated with the emitted rays is always less than the energy associated with the excitation rays because the excited molecules tend to lose some of their energy ($\Delta h\nu$) by interaction with the impurities (salt ions, cellular components, particulate matter, etc.) in the solution or tissue.

3.1 Labeling Antibodies with Fluorochromes

Antibodies can be labeled by direct coupling to fluorochromes. For example, detection of $CD4^+$ Th (T-helper cells) in a human blood film can be performed by using an anti-human $CD4^+$ antibody bound to fluorescein. Thus detection of the target cell(s) by using a specific fluorochrome-tagged antibody is called direct fluorescence. Alternatively, at a first instance, an anti-human $CD4^+$ antibody (without fluorochrome molecule) can be used followed by an anti-human antibody tagged with a fluorochrome (say a secondary antibody, rabbit anti-human IgG-fluorescein). This approach is called indirect fluorescence. For indirect work, suitable secondary reagents labeled with fluorochromes are available commercially. Direct labeling may be the method of choice where such conjugates are not available or where a direct label is required specifically, for example, in simultaneous visualization of two antibodies of the same class or subclass. The most commonly used fluorochromes are fluorescein and rhodamine. They can be conjugated to anti-immunoglobulin antibodies, protein A, protein G, avidin, or streptavidin. These conjugates are available from many commercial sources or can be prepared in the laboratory.

3.2 Detection of Fluorochrome-Labeled Reagents

To detect fluorochrome-labeled reagents, a specially equipped microscope is required. The low levels of fluorescence produced in cell staining experiments mean that the microscope must be equipped for epifluorescence in which the exciting radiation is transmitted through the objective lens onto the surface of the specimen. Absorbing radiation of the appropriate wavelength causes the electrons of the fluorochrome to be raised to a higher energy level. As these electrons return to their ground state, the light of a characteristic wavelength is emitted. This emitted light forms the fluorescent image seen in the microscope. Individual fluorochromes have discrete and characteristic excitation and emission spectra. Filters are used to ensure that the specimen is irradiated only with light at the correct wavelength for excitation. By placing a second set of filters in the viewing light path that only transmits light of the wavelength emitted by the fluorochrome, images are formed only by the emitted light. This produces a black background and a high-resolution image.

Because some fluorochromes have emission spectra that do not overlap, two fluorochromes can be observed on the same sample. This allows the study of two different antigens in the same specimen even when they have identical subcellular distribution. However, fluorescence detection is not compatible with most histochemical stains, because the components of most of these stains auto-fluoresce strongly. Fluorescence detection is also not compatible with enzyme detection systems, because the deposition of insoluble compounds after enzyme detection will block the emission of light from the fluorochrome.

An extension of the immuno-fluorescence technique is fluorescence-activated cell sorter (FACS), which is capable of detecting, counting, or sorting various cell types on the basis of their size (30–100 micrometer size), viability (using propidium iodide that binds DNA of dead cells), granularity, and specific surface receptors. Thus FACS uses multi-parametric analyses for detection, counting, and sorting of a specific type of cells of interest. The FACS is used for quantitative detection and analysis of T-helper cells (Th cells; CD4$^+$) and cytotoxic T cell (Tc cells; CD8$^+$) to monitor the status of AIDS patients and to study the effectiveness of the anti-AIDS therapy.

3.3 Choosing the Correct Fluorochrome

The choice of fluorochrome is limited primarily by filter sets that are commercially available for the microscope. Most filter sets are best matched to the properties of rhodamine or fluorescein. Texas Red can be used with rhodamine filter sets, but these filters do not exactly match its emission spectrum. The increasing availability of phycobiliproteins, which are theoretically about 50 times brighter, is going to have a substantial influence on the design of immuno-fluorescence experiments over the next few years as other filter sets become available. Fluorescein emits a yellow-green light that is detected well by the human eye and by most films. However, fluorescein is prone to rapid photobleaching, and bleaching retardants such as DABCO or o-phenylenediamine should be added to the mounting medium. Rhodamine emits a red color and is not as prone to fading as fluorescein, but the rhodamine conjugates are more hydrophobic and therefore yield higher backgrounds than fluorescein. Texas Red also emits a strong red light and its emission spectrum is different from the emissions of fluorescein than rhodamine. It is the least likely to produce problems of fading, but it is not as widely available as either fluorescein or rhodamine. While using dual labels, fluorescein can be combined with either Texas Red or rhodamine [1, 2].

4   Immunoassay: Enzyme-Linked Immunosorbent Assay (ELISA)

Immunoassays are among the most powerful, sensitive, and specific immunochemical techniques. In ELISA either antigen or antibody is initially bound to the surface provided by the wells of a microtiter plate and antibody or antigen is detected by the following steps. The ELISA employs a wide range of methods to detect and quantitate antigens or antibodies. This method has been successfully exploited to detect viruses, bacteria, fungi, and mycoplasma by detecting the presence of surface-exposed antigens on the surface of these microbes or indirectly by determining the presence of pathogen-/antigen-specific antibodies in the serum or other body fluids. Also, ELISA has been used to detect the presence of various hormones such as hCG, progesterone, FSH, LH, insulin, glucagon, thyronine, prolactin, etc. besides a range of cytokines including interleukins, interferons, etc. There are many variations in which immunoassay can be performed, and they are classified on the basis of many different criteria. Within each type, the principle and the order of the steps are similar. However, the variable that is being tested may change. By changing certain key steps, an assay can be altered to determine either antigen or antibody level. The ELISA can be classified into three broader types:

(i)   Antibody-capture (competitive) ELISA

(ii) Two-antibody sandwich ELISA

(iii) Antigen-capture ELISA

All three types can be performed by the direct or indirect approach. In direct ELISA, the probe antibody is directly conjugated to an enzyme, while in the case of indirect approach a second anti-species antibody is conjugated to an enzyme. The enzymes, which are covalently conjugated to the antibody, possess high turnover number, are highly stable, and on reaction with a suitable chromogenic substrate result in the formation of soluble or insoluble colored end product. By measuring the absorbance at the particular wavelength, the intensity of the color is read. The higher the absorbance, the more shall be the concentration of the antigen or antibody being detected. In a qualitative assay, the development of color is indicative of a positive reaction. Another variation of ELISA that uses properties of both direct and indirect detection is the biotin–streptavidin system. Here, the antigen or antibody is purified and labeled with biotin. The biotinylated reagent is detected by binding with streptavidin that has been labeled with an enzyme.

In an antibody-capture assay, the antigen is attached to a solid support, and a labeled antibody is allowed to bind. After washing, the assay is quantitative by measuring the amount of antibody retained on the solid support. In an antigen-capture assay, the antibody is attached to a solid support and labeled antigen is allowed to bind. The unbound proteins are removed by washing, and measuring the amount of bound antigen quantitates the assay [1, 2].

4.1 Antibody-Capture Immunoassay

An indirect ELISA is used to measure an antigen-specific antibody level in the samples. A putative antiserum (may contain antigen-specific antibody) is reacted with a specific antigen attached to a solid phase. The specific antibody molecules bind to the antigen and all other materials are washed away. Exposure of the antigen–antibody complex to an enzyme-labeled anti-immunoglobulin antibody results in binding to specific antibody molecules adsorbed from the serum sample. The complex is washed and the substrate for the enzyme is added, resulting in the formation of a colored end product whose concentration is directly proportional to the amount of specific antibody in the serum sample under analysis (Fig. 5). This type of ELISA is used to detect a pre-exposure to hepatitis C or hepatitis E virus by measuring the level of virus-specific IgG-antibody in the serum of the suspected patient.

ELISA has replaced radioimmuno assay (RIA) despite the latter being extensively automated and even more sensitive. This is because ELISA is cheaper, lacks the radiological hazards of RIA, and is suitable for use in small laboratories lacking radioactivity-counting facilities.

A variation of this method is called competitive ELISA. In this method, a mixture of a known amount of enzyme-labeled antigen and an unknown amount of unlabeled antigen is allowed to react with a specific antibody attached to a solid phase. After the complex has been washed with buffer, the enzyme substrate is added and the

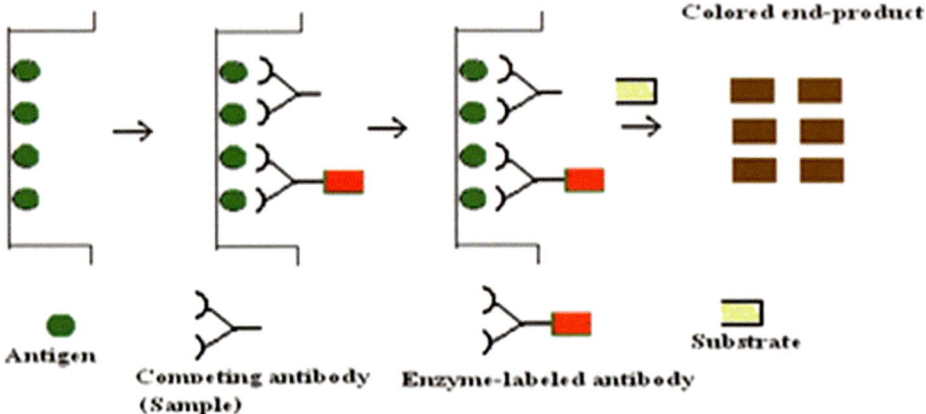

Fig. 5 Detection of antigen-specific (unlabeled competing) antibody by antibody-capture ELISA

enzyme activity is measured. The difference between this value and that of a sample lacking unlabeled antigen is a measure of the concentration of unlabeled antigen. A major limitation of this technique is that each antigen may require a different method to couple it to the enzyme. This limitation is overcome in two-antibody sandwich ELISA.

4.2 Two-Antibody Sandwich ELISA

In the two-antibody sandwich assay, one antibody specific for a particular antigen is permitted to bind on the solid surface, and the antigen (infected sample like serum, blood, CSF, urine, saliva, etc.) is allowed to bind to the first antibody. The second antibody that is also specific for the same antigen but conjugated to an enzyme is added. The unbound second antibody is washed off with a buffer solution and the enzyme substrate is added. Measuring the amount of chromogenic end product produced by the enzyme-labeled second antibody, which is also specific to the same antigen and hence binds it, makes it a quantitative assay. The amount of colored product produced measured under standard conditions is directly proportional to the amount of antigen present in the sample (Fig. 6). The two antibodies specific for an antigen used in this assay are raised in two different animals. For example, while the first antibody is raised in rabbit against an antigen "X," the second antibody is produced in mouse (say a monoclonal antibody) using the same antigen ("X"). This type of ELISA is used in the detection of hepatitis A virus and hepatitis B virus surface antigen (HBsAg) to confirm infection. To detect and quantitate antigens, the most useful method is the two-antibody sandwich assay. These assays are quick and reliable and can be used to determine the relative levels of most protein antigens.

The sensitivity of any type of ELISA may be greatly enhanced by enzyme amplification. The primary enzyme product is used to trigger a secondary coupled enzyme system that can generate a

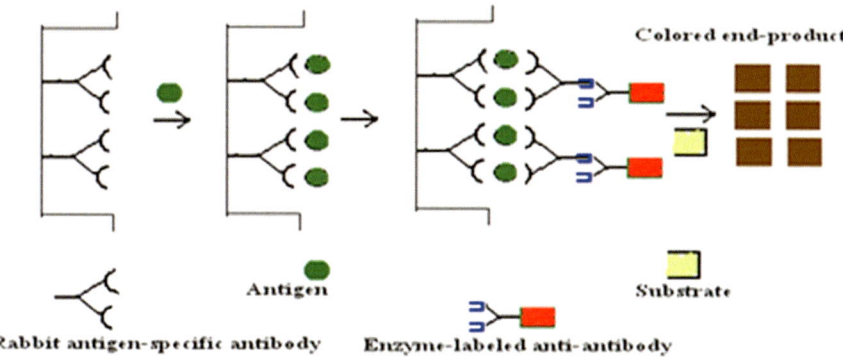

Fig. 6 Two-antibody sandwich ELISA for detection of specific antigen

large quantity of chromogenic product. Since the product of the first enzyme need not be measured but acts catalytically only on the second system, enzymes not currently used for ELISA may become important in these systems, e.g., aldolase or glucose-6-phosphatase. Enzyme amplification of a double-antibody assay in which alkaline phosphatase, the primary enzyme, degrades the substrate $NADP^+$ to NAD can be achieved by using alcohol dehydrogenase as the amplifying enzyme and a tetrazolium dye as a redox acceptor. When alcohol dehydrogenase is triggered, the tetrazolium dye is reduced to a colored formazan product, which can be assayed spectrophotometrically; e.g., iodonitrotetrazolium violet may be reduced to a red formazan and yellow thiazolyl blue may be reduced to a blue formazan. Under appropriate conditions, 500 molecules of formazan are produced by the primary enzyme system. Enzyme amplification can be achieved as a one-step process in which both enzymes and substrates are reacting at the same time or a two-step amplification in which the first enzyme is inhibited before or during the addition of the second enzyme and substrate. Enzyme amplification immunoassays have already been developed for hormones, viruses, bacteria, and tumor markers.

4.3   Antigen-Capture Assay

In this assay, the antibody is immobilized to the solid surface. An enzyme-conjugated antibody is then added followed by extensive washings with the buffer to wash off the free enzyme-conjugated antibody. Subsequently, a suitable substrate specific for the enzyme is added and the colored end product is measured spectrophotometrically (Fig. 7). This type of assay is primarily used to detect and quantitate antigens in a given sample. The major limitation of this assay is that each of the different antigens must be individually labeled with an enzyme. This approach of making various enzyme-conjugated antigens is laborious and costly. Thus, the two-antibody sandwich assay is most commonly used to perform the antigen-capture assay with very high sensitivity and specificity.

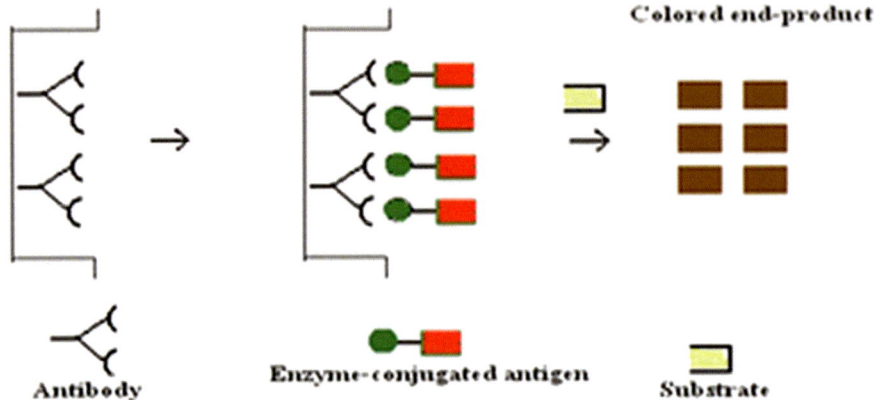

Fig. 7 The concept of antigen-capture immunoassay

4.4 Antibody Detection by Indirect ELISA

Antigen-specific antibody detection by indirect ELISA can be used to detect and quantitate antibodies and to compare the epitopes recognized by different antibodies. When labeled antibody assays are performed with excess antigens on the solid phase (i.e., enough to saturate all the available antibodies), the presence and level of antibodies in a test solution can be measured (Fig. 8). The general protocol is simple; an unlabeled antigen (purified or partially purified) is immobilized on a solid phase, and the antibody (in a body fluid sample, e.g., blood, serum, cerebral spinal fluid, saliva, etc.) is allowed to bind to the immobilized antigen. The antibody can be labeled directly or can be detected by using a labeled secondary anti-antibody that will specifically recognize the antibody. The amount of antibody that is bound determines the strength of the signal.

4.5 Detection

All immunoassays rely on labeled antigens, antibodies, or secondary reagents for detection. Three factors that will affect the sensitivity of a labeled antibody assay are (1) the amount of antigen that is bound to the solid phase, (2) the avidity of labeled moieties used to label the antibody, and (3) the type and number of labeled moieties used to label the antibody. Antibodies usually are labeled with enzymes or biotin. These can also be labeled with radioactive compound and fluorochromes. Of the methods used to label, radioactivity has biological hazards. The fluorochromes have few applications for immunoassays. Although they can be very sensitive, they require expensive equipment to use and, unlike radiometric or enzymatic detection, no alternative methods can be used to locate and quantitate positives. Various enzymes can be used to label antigens or antibodies such as horseradish peroxidase (HRP) or alkaline phosphatase (AP). These enzymes act on suitable substrates to yield soluble or insoluble end product (Table 2). Some chromogenic substrates may be carcinogenic and shall be handled with utmost care.

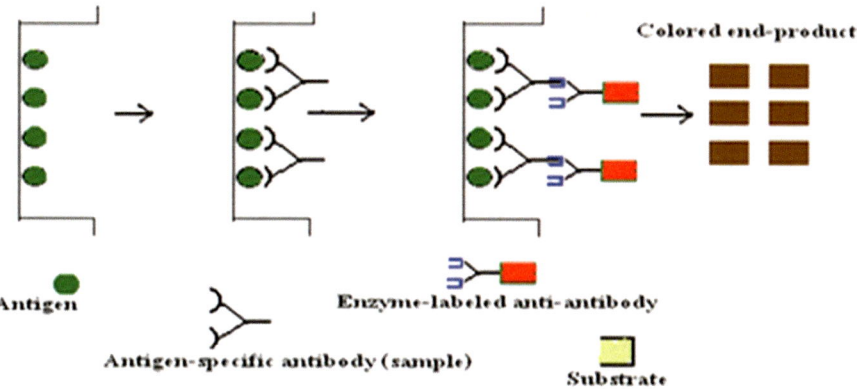

Fig. 8 Detection of specific antibody by indirect-ELISA

Table 2
Various combinations of enzymes and chromogens

Enzyme	Substrate	End product
o-Horseradish peroxidase (HRP)	Phenylenediamine (OPD)	Soluble
	Tetramethylbenzidine (TMB)	Soluble
	4-Chloro-1-naphthol (CN)	Insoluble
	3-Amino-9-ethylcarbazole (AEC)	Insoluble
	3, 3′, 4, 4′-Tetraaminobiphenyl (DAB)	Insoluble
p-Ni alkaline phosphatase (AP)	Triphenyl phosphate (PNPP)	Soluble
	Bromochloroindolyl phosphate-nitro blue tetrazolium (BCIP/NBT)	Insoluble

5   Radioimmunoassay (RIA)

RIA is a nuclear technique widely used for measuring minute substances with in vitro procedures. It combines the technology of nuclear medicine (tracer technique) and immunology (antigen-antibody binding) designating the name RIA. Someone even called it a hybrid technique. This technique has been developed in 1959 by Solomon Berson and Rosalyn Yalow in Bronx, New York. For her contributions to this important analytical technique of medical science, R. Yalow shared the 1977 Nobel Prize in Medicine and Physiology. In earlier days RIA is only limited to a certain hormone, but now the scope has greatly expanded to the field of reproductive physiology, oncology, immunology, hematology, pharmacology, parasitology, etc. It makes a great contribution to the diagnostic laboratory and scientific research works.

Principle:

Ag, Ab, and *Ag reactions

Ag – Antigen to be assayed in serum (unknown or standard)

Variant*Ag – Labeled antigen added in a minute amount of radio-activity equally in each tube

Nonvariant (constant) Ab – Antibody added in a minute amount of dilution equally in each tube

Nonvariant (constant) and limited

Based on competition between unlabeled antigen and a finite amount of corresponding labeled antigen for a limited number of antibody binding sites in a fixed amount of antiserum. At equilibrium in the presence of an antigen excess, there will be both the free antigen and the antigen bound to the antibody. Competing depression reaction labeled antigen (*Ag) possesses the same properties as an unlabeled antigen (Ag). It can also bind to the correlated specific antibody (Ab) with the formation of a labeled antigen–antibody complex (called bound antigen (B)), leaving the unbound one as a free labeled antigen (F). The more Ag is present, the less likely is the *Ag bound to the Ab; thus the amount of B formed is inversely proportional to the Ag originally present in serum; this is the so-called competing depression reaction.

- $4Ag^* + 4\,Ab \rightarrow 4Ag^* \cdot Ab$
- $4\,Ag + 4\,Ag^* + 4\,Ab \rightarrow 2Ag^*Ab + 2\,AgAb + 2Ag^* + 2Ab$
- $12\,Ag + 4\,Ag^* + 4\,Ab \rightarrow Ag^*Ab + 3AgAb + 3Ag^* + 9\,Ag$

With unbound Ag* and Ag washed out, the radioactivity of bound residue is measured. Ligand concentration is inversely related to radioactivity [Ag, ligand to be measured; Ag*, radiolabeled ligand].

RIA Tracers:

1. *Internally labeled molecules*

Typically, *tritium (3H) is the label, sometimes 14C.* Usually purchased commercially. Used only for small molecules like steroids or drugs. Pro: the tracer immunologically behaves exactly as the cold hormone, thus theoretically perfect. Con: beta radiation is weak and therefore more difficult to measure, thus practically cumbersome. Beta rays have low penetrating power: the radioactive sample needs to be mixed with a scintillator fluid; the produced light is measured by use of a photomultiplier ("beta counter") [1, 2].

2. *Externally labeled molecules*

Typically, *125I is the label*. Pros: it is often produced in the research lab itself and gamma radiation has high penetrating power; it is therefore easy to measure, thus practical to use. Con: the tracer immunologically does *not always* behave exactly as the cold hormone, due to iodination damage.

Con: shelf life of iodinated protein is < 4 weeks. Usually, 125I will take the place of a hydrogen atom on the ring of tyrosine. Sometimes, a radioactively labeled molecule needs to be conjugated to the protein.

To perform RIA, a quantity of an antigen is made radioactive by labeling it with gamma-radioactive isotopes of iodine attached to amino acid tyrosine on it. The radioactive isotope I^{125} I^{131} is used. The radiolabeled antigen is then mixed with a known amount of antibody present in an RIA tube for that antigen, and then a sample of serum from a patient containing an unknown quantity of the same antigen is added. This causes the competition between unlabeled antigen and a labeled antigen for antibody binding sites. If the concentration of unlabeled antigen is increased, the more it binds to the antibody, displacing radiolabeled antigen. The bound antigens are separated from unbound ones as shown in Fig. 9.

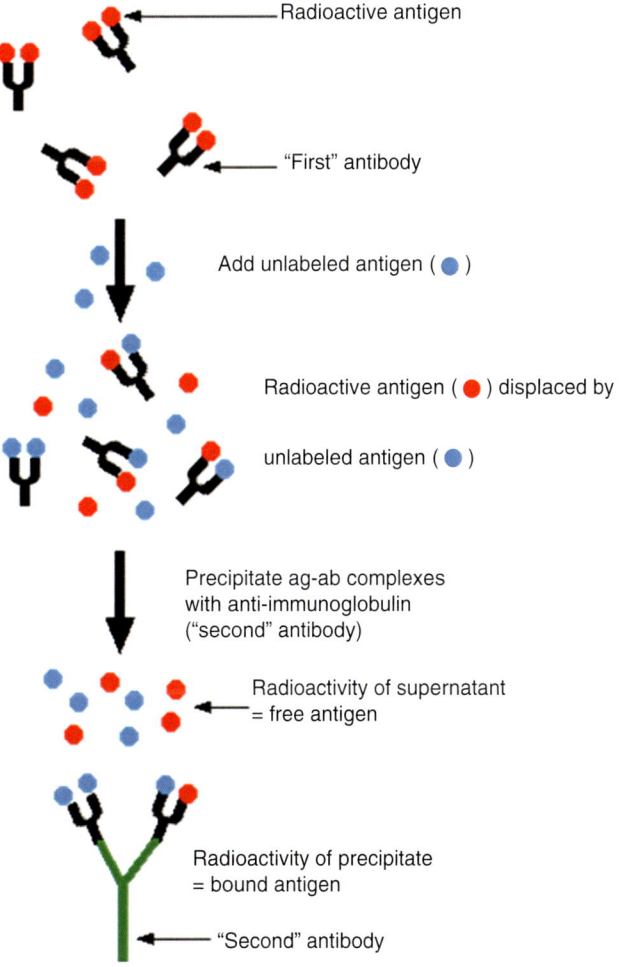

Fig. 9 Radioimmunoassay

- The bound counted in pellet form.
- The free counted in the supernatant fluid.
- The radioactive free antigen remaining in the supernatant is measured using a gamma counter.
- A gamma counter is a machine used to measure gamma radiation emitted by isotopes.

Gamma Counters Used in RIA

- *Standard substance*: The fundamentals of accurate quantitation – they must be same as the antigen which will be assayed.
- *Labeled antigen*

  1. Minute amount: less than the minimum of the Ag to be assayed
  2. Optimal specific activity
  3. Radiochemical purity :>95%
  4. Good immunoactivity
  5. Stability

- Separating reagent (SR). Separation of B and F:

  1. Non-specific SR: Absorbing F
  2. Specific SR: Absorbing B, a second antibody
  3. Solid-phase SR: Absorbing Ab on solid material

---

6   Western Blotting or Immunoblotting

Immunoblotting combines the resolution of gel electrophoresis with the specificity of immunochemical detection. Immunoblotting can be used to determine the number of important characteristics of protein antigens, the presence and quantity of an antigen, the relative molecular mass of the polypeptide chain, and the efficiency of extraction of the antigen. It is particularly useful when dealing with antigens that are insoluble, difficult to label, or easily degraded and thus less amenable to analysis by immunoprecipitation. Immunoblotting can be combined with immunoprecipitation to permit very sensitive detection of minor antigens and to study the specific interaction between antigens. It is also a particularly powerful technique for assaying the presence, quantity, and specificity of antibodies from different samples of polyclonal sera. Moreover, it can be used to purify specific antibodies from polyclonal sera. Because the antigen is not labeled, only the steady-state level can be determined, and further analysis of its biochemical properties, modifications, or half-life is not possible [1, 2].

The immunoblotting procedure can be divided into the following six steps:

(i)   Preparation of antigen sample

(ii) Resolution of the sample by SDS-PAGE

(iii) Transfer of the separated polypeptides to a membrane support

(iv) Blocking of non-specific sites on the membrane

(v)   Addition of the probe (enzyme, radioactivity, or fluorochrome-labeled antibody)

(vi) Detection

First, an unlabeled solution of proteins, for example, an extract of cells or tissue homogenate, is prepared in a gel electrophoresis sample buffer. The proteins are separated by gel electrophoresis and transferred to a nitrocellulose membrane that binds the proteins non-specifically. The transfer usually is achieved by placing the membrane in direct contact with the gel and then placing the sandwich in an electric field to electrophoretically move the proteins from the porous gel onto the membrane (Figs. 10 and 11). The remaining binding sites on the membrane are blocked to eliminate any further reaction with the membrane (Figs. 10 and 11). Finally, the location of specific antigens is determined using a labeled primary antibody or an unlabeled primary antibody, followed by a labeled secondary antibody. The antibody is often labeled with alkaline phosphatase or horseradish peroxidase. Addition of a suitable chromogen and the substrate specific for the

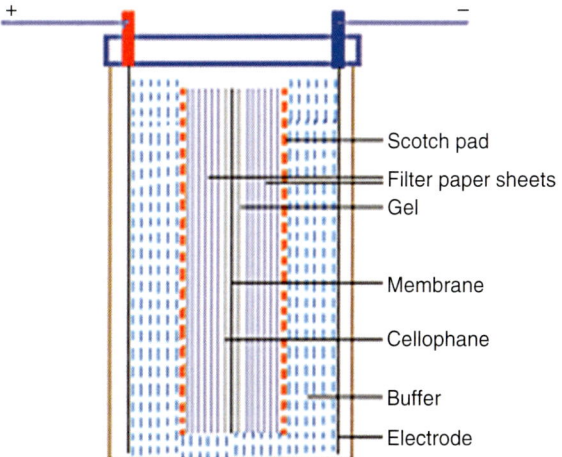

Fig. 10 Diagrammatic representation of a vertical electroblotting set up. This technique is also known as wet-electroblotting. The proteins already resolved on polyacrylamide gel are transferred onto the membrane under the influence of electric fiels, and their passage through the pores of the membrane is avoided by using a cellophane membrane (molecular cut-off limit of 10 kDa) underneath the nitrocellulose membrane

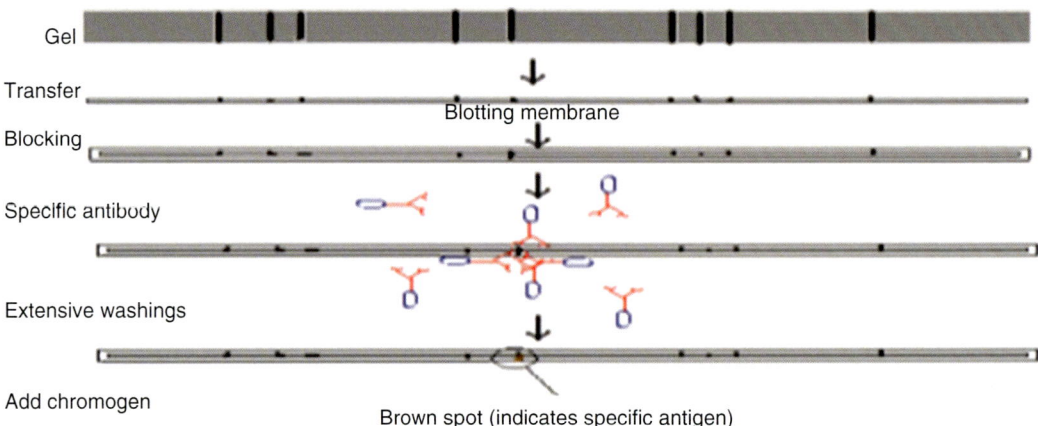

Fig. 11 Important steps in the localization of specific antigen(s) during Western blotting

enzyme conjugated to antibody results in the formation of colored insoluble precipitates (end product) on the membrane indicating presence of a particular antigen that is being recognized by the labeled antibody.

Transfer of proteins from the gel to nitrocellulose membrane can be achieved in one of the two ways. In capillary blotting, the gel is placed on a wet pad of buffer-soaked filter paper and a sheet of membrane placed on the gel. The buffer is drawn through the gel by a heavy weight on top of the membrane sheet. Passage of buffer by capillary action through the gel transfers the separated proteins onto the membrane, to which they bind irreversibly by hydrophobic interaction. The process of capillary blotting is often carried out for 16–24 h that permits transfer of proteins from the gel onto the membrane. Thus a relatively small amount (10–20%) of each protein in the gel is transferred onto the membrane. A quicker and more efficient method involves the application of electric field to hasten the transfer of proteins from the gel onto the membrane and is referred to as electroblotting (Fig. 10).

A sandwich of gel and nitrocellulose membrane soaked in transfer buffer is underlaid and overlaid with blotting sheets cut to the size of the gel. This sandwich is compressed in a cassette and immersed in the buffer, between two parallel electrodes. The current is passed at a right angle to the gel that causes the electrophoretically resolved proteins to migrate out of the body of the gel and get transferred (blotted) on the membrane. The membrane with the transferred proteins can be examined further by the specific antibody as mentioned above.

In a variation of the wet electroblotting, a semi-dry electroblotting system is preferred that uses only a limited volume of the transfer buffer and in which only a couple of sheets of filter paper are soaked, making it economical. A cassette comprising gel and nitrocellulose membrane soaked in transfer buffer is underlaid and overlaid with filter paper sheets cut to the size of the gel. This cassette is placed on a wet graphite anode. The uppermost layer of the cassette comprising buffer-soaked blotting sheets is contacted with a wet graphite cathode. Graphite is the best material for electrodes in semi-dry blotting because it conducts well, does not overheat, and does not catalyze oxidation products. A current not higher than 0.8–1 mA per cm^2 of blotting surface is recommended. The gel can overheat if higher currents are used and proteins can precipitate. The protective casing is closed and current is applied across the pair of electrodes. The proteins in the gel under the influence of applied electric field migrate out of the gel and get electroblotted on the membrane underneath in 60–90 min. The larger proteins take more time to transfer out of the gel than the smaller ones. This transfer time is required to be optimized for different proteins and for the gels of varying thickness. The membrane is subsequently examined by use of a specific probe antibody (Fig. 11).

6.1 Major Limitations

The major factor that will determine the success of an immunoblotting procedure is the nature of the epitopes recognized by the antibodies. The gel electrophoresis technique involves denaturation of the antigen sample, so only antibodies that recognize denaturation-resistant epitopes will bind the antigen. Most polyclonal sera contain at least some antibodies of this type, but many monoclonal antibodies will not react with denatured antigens.

The partial or incomplete transfer of the electrophoretically resolved proteins from the gel onto the membrane also limits the detection of antigens by the probe antibody. Generally, the high molecular weight proteins/antigens take more time to transfer than the relatively low molecular weight antigens. Thus the transfer time for optimal electroblotting of antigens onto the membrane shall be optimized for different antigen mixtures.

The antigen density immobilized at the surface of the membrane also favors the retention of low-affinity antibodies on the membrane by increasing the frequency with which antibodies leaving the membrane bind to adjacent sites. The sensitivity of immunoblotting is also determined by the detection method. For most detection systems, this limit will be about 20 femtomoles. For a 50-kDa protein, it shall be approximately 1 ng. The loading capacity of the gels is limited and an antigen usually cannot be detected when its concentration falls below 1 ng/sample. In case of a typical SDS-PAGE, the capacity of a lane is approximately 150 μg. Distortion is seen if large amounts are used. The detection limit of most immunoblots is 1–10 ng.

6.2   Choice of Probe Antibody

Three types of antibody preparations can be used for immunoblotting, which include polyclonal antibodies, monoclonal antibodies, and pooled monoclonal antibodies.

7   Southern Blotting

In the 1970s Ed Southern of Oxford University invented a revolutionary DNA blotting technique. The Southern blot allows the visualization of one DNA fragment from a whole genome DNA extract.

Concept: *Reannealing nucleic acids to identify the sequence of interest. It separates DNA/RNA in an agarose gel and then detects specific bands using probe and hybridization. Hybridization takes advantage of the ability of a single-stranded DNA or RNA molecule to find its complement, even in the presence of large amounts of unrelated DNA. It allows detection of specific bands (DNA fragments or RNA molecules) that have a complementary sequence to the probe. It indicates size of bands and quantifies the abundance of a molecule ([1]; [2]).*

Southern Blot: DNA–DNA*:
It uses gel electrophoresis together with hybridization probes to characterize restriction fragments of genomic DNA (or DNA from other sources, such as plasmids).

- It identifies DNA with a specific base sequence.
- It can be done to detect specific genes present in cells.

Step 1

1. DNA to be analyzed is digested to completion with a restriction endonuclease.
2. Electrophoresis to maximally separate restriction fragments in the expected size range. A set of standards of known size is run in one lane of the gel.
3. Blot fragments onto a nitrocellulose membrane.
4. Hybridize with the ^{32}P probe.
5. Autoradiography.

Step 2
- Separate DNA fragments.
- *Soak gel in 0.5 M NaOH.*
- Convert dsDNA to ssDNA

Step 3
- Cover gel with nitrocellulose paper.
- Cover nitrocellulose paper with a thick layer of paper towels.

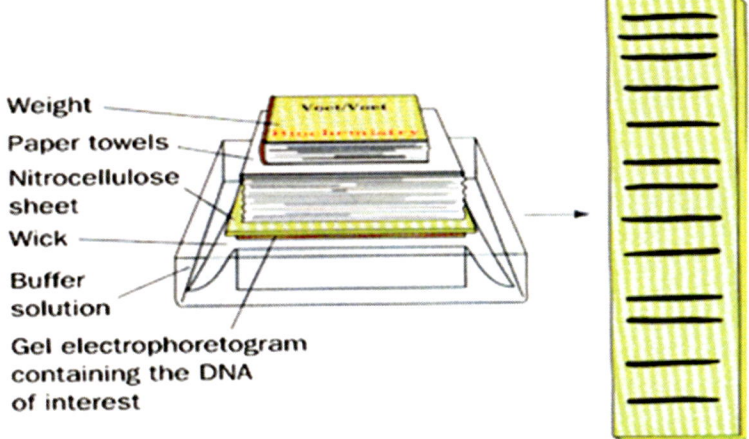

Fig. 12 Vacuum dry nitrocellulose at 80 °C to permanently fix DNA in place or crosslink (via covalent bonds) the DNA to the membrane

- Compress apparatus with heavy weight.
- ssDNA binds to nitrocellulose at the same position it had on the gel.
- Vacuum dry nitrocellulose at 80 °C to permanently fix DNA in place or cross-link (via covalent bonds) the DNA to the membrane as shown in Fig. 12.

Step 4
- Incubate nitrocellulose sheet with a minimal quantity of solution containing ^{32}P-labeled ssDNA probe.
- Probe sequence is complementary to the DNA of interest.
- Incubate for several hours at suitable renaturation temperature that will permit the probe to anneal to its target sequence (s).
- Wash and dry nitrocellulose sheet as shown in Fig. 13.
  Southern Application:
  Diagnosis and detection of genetic diseases

- Used to diagnose sickle cell anemia.
- A→T base change in the β subunit of Hb:
  - Glu→ Val
- Development of a 19-residue oligonucleotide probe complementary to sickle-cell gene's mutated segment.
- The probe hybridizes to DNA from homozygotes of sickle-cell anemia but not from normal individuals.

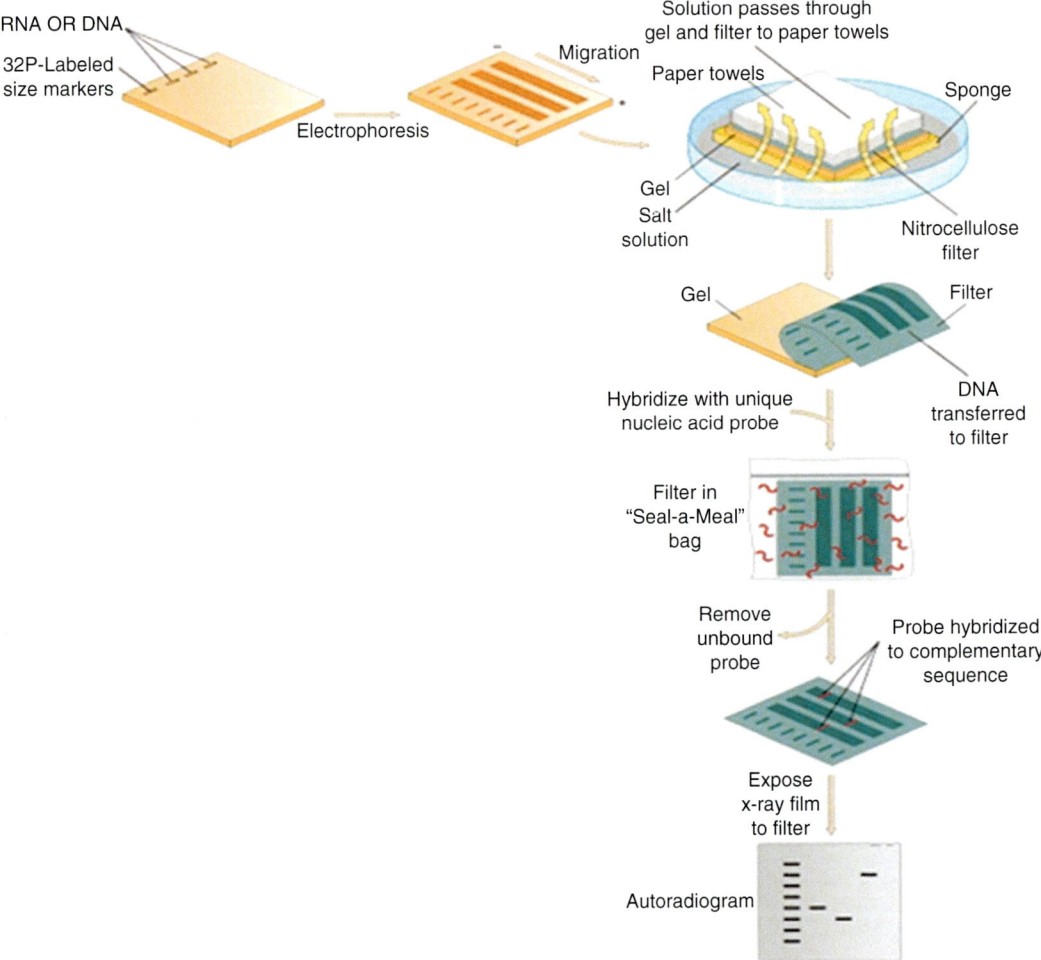

Fig. 13 General scheme for Southern blotting

8   Northern Blot: RNA–DNA*(RNA*)

Alwine adapted Southern's method for DNA to detect size and quantify RNA (1977). No need to digest RNA with restriction enzymes. Use formaldehyde to break H-bonds and denature RNA because single-stranded RNA will form intramolecular base pairs and "fold" on itself.

Isolate RNA and treat with formaldehyde. Electrophorese RNA in denaturing agarose gel (has formaldehyde). Visualize RNA in the gel using ethidium bromide stain and photograph. Transfer single-stranded RNA to nitrocellulose or nylon membrane. Covalently link RNA to the membrane. Incubate membrane (RNA immobilized on the membrane) with labeled DNA or RNA probe with the target sequence. And the last step is development of autoradiography.

Step 1:

Isolate RNA:

- To detect rare mRNA, isolate the poly A$^+$ mRNA.
- RNA is both biologically and chemically more labile than DNA. Thus eliminate RNases.

Step 2:

Electrophoresis:

- Performed in formaldehyde agarose gel to prevent RNA from folding on itself.
- Stain with EtBr to visualize the RNA bands.

Step 3:

Transfer single-stranded RNA to nitrocellulose or nylon membrane:

Traditionally, a nitrocellulose membrane is used, although nylon or a positively charged nylon membrane may be used. Nitrocellulose typically has a binding capacity of about 100 μg/cm, while nylon has a binding capacity of about 500 μg/cm. Many scientists feel nylon is better since it binds more and is less fragile.

Covalently link RNA to membrane: UV cross-linking is more effective in binding RNA to the membrane than baking at 80 °C.

Steps 4 and 5:

Prehybridize before hybridization: Blocks non-specific sites to prevent the single-stranded probe from binding just anywhere on the membrane.

-Incubate membrane with labeled DNA or RNA probe with target sequence:

The probe could be ^{32}P, biotin/streptavidin, or a bioluminescent probe.

Autoradiography::

Place membrane over X-ray film.

X-ray film darkens where the fragments are complementary to the radioactive probes.

Northern Application:

Northern blots are particularly useful for determining the conditions under which specific genes are being expressed, including which tissues in a complex organism express genes at the mRNA level.

For instance: When trying to learn about the function of a certain protein, it is sometimes used to purify mRNA from many different tissues or cell types and then prepare a Northern blot of those mRNAs, using a cDNA clone of the protein of interest as the probe.

The only mRNA from the cell types that are synthesizing the protein will hybridize to the probe.

Difference Between Southern and Northern Blotting:

<u>Southern</u>	<u>Northern</u>
● DNA on membrane.	● RNA on membrane.
● Digest DNA.	● No need to digest DNA.
● Convert dsDNA to ssDNA.	● Denature "folded" RNA with formaldehyde.
● Probe with DNA or RNA.	● Probe with DNA or RNA.

References

1. Rajan K (2011) Analytical techniques in biochemistry and molecular biology. Springer, New York. eBook ISBN978-1-4419-9785-2

2. Wilson K, Walker J (2010) Principles and techniques of biochemistry and molecular biology. Cambridge University Press, Cambridge. ISBN 978-0-521-73167-6

Chapter 8

Radioactivity

Abstract

In chemistry and biochemistry, we are used to chemical reactions where one compound is turned into another. We can identify and measure ("assay") the reactants and products and learn something about the reaction. To study this we need some method for detecting the product of the reaction and this is often done with isotopes. Why do we need a radioisotope? We now need to understand what radioactivity is and how to use it.

Key words Radioactivity, Ionization chambers, Proportional counters, Liquid scintillation counting, Solid scintillation counting, Tracer technique

Abbreviations

PET   Positron emission tomography
SPECT   Single-photon emission computed tomography

Nuclear decay or *radioactivity* is the process by which a nucleus of an unstable atom loses energy by emitting ionizing radiation.

A material is said to be radioactive if it spontaneously emits this kind of radiation which includes the emission of alpha particles, beta particles, gamma rays, and conversion electrons.

Unstable nucleus: It has excess energy, wants to go to "ground state," and becomes stable by emitting ionizing radiation.

Mahin Basha, *Analytical Techniques in Biochemistry*, Springer Protocols Handbooks,
https://doi.org/10.1007/978-1-0716-0134-1_8, © Springer Science+Business Media, LLC, part of Springer Nature 2020

1   Radiation Types

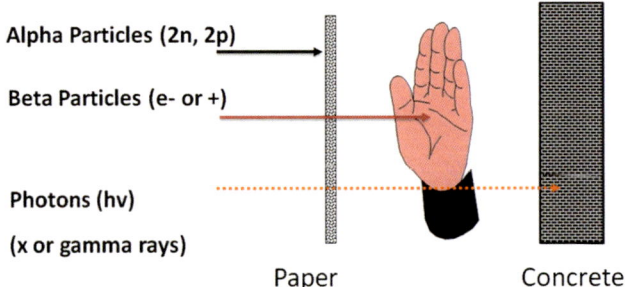

Half-Life and Mean Life:
Half-life is the time required for half of the atoms of a radioactive material to decay to another nuclear form.
Mean life is the average of all half-lives.

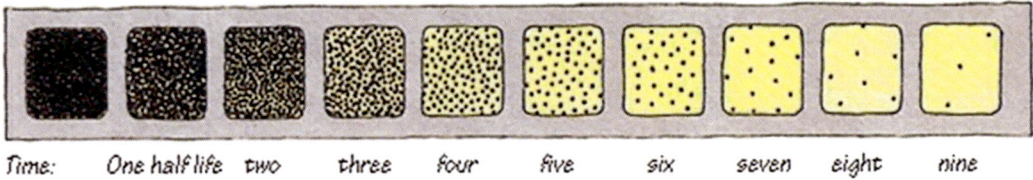

Decay rate of radioactivity: After ten half lives, the level of radiation is reduced to one thousandth

2   Measurement of Radioactivity

The three commonly used methods are:

1. Ionization of gases
2. Excitation of solids and liquids
3. Ability to expose photographic emulsions, i.e., autoradiography

2.1   Ionization of Gases

Principle:
Radiation causes ionization of gaseous particles in its path. When this takes place between electrodes in a closed chamber, an electric pulse is generated. The magnitude of it depends upon applied potential and number of radiation particles entering the chamber.

Instrumentation:
Ionization chamber consists of a gas chamber with insulated base. It consists of an anode wire and metallic cathode. The circuit is closed by connecting to a battery and resistor. The electric pulse produced is measured by amperometer as shown in Fig. 1 [1].

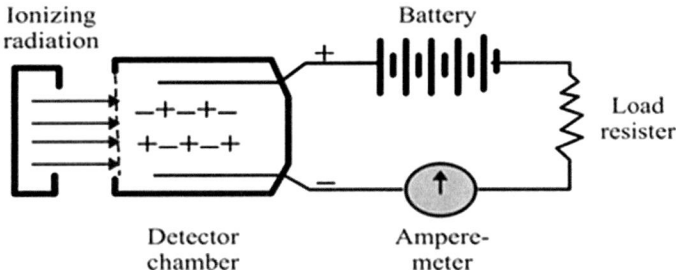

Fig. 1 Key component in a simple ionization chamber

3 Types

3.1 Ionization Chambers or Simple Ionization

In this type only one ion pair is produced per collision.

Disadvantages:

- Current production is low.
- Sensitive devices are needed for detection.

3.2 Proportional Counters or Gas Amplification

At high voltage ionized electrons move with high speed. This causes secondary and tertiary ionization. Finally, a large number of electrons reach the anode. This is known as the Townsend avalanche effect.

Advantages:

- Current production is high.

Disadvantages:

- Constant voltage is needed as it affects ionization and amplification.

Uses:

- To detect alpha-emitting isotopes

3.3 Geiger–Muller Counters

All the isotopes including beta emitters induce complete ionization. Current flow is independent of primary ions. Maximal gas amplification is seen. The output pulse is same in considerable voltage range. The number of times the pulse is produced will give the radioactivity. Dead time is the time taken by the ion pairs to reach their respective electrodes. This is usually 100–200 μS.

Advantages:

- Can detect beta radiation

Disadvantages:

- Ionizing particles entering into the tube during the dead time cannot be detected.
- Some ions escape from the electrodes and give continuous discharge.

Uses:

- Routine check of radioactive laboratory for contamination
- Qualitative analysis of radioactivity
- Quick screening of gels and chromatographic fractions for labeled components

---

4   Excitation of Solids and Liquids

Principle: *Radioactive isotopes can excite a compound called fluor. This emits photons of light which can be detected and quantified using a photomultiplier tube. This is called scintillation counting.*

Types – There are two types of scintillation counting:

1. Solid scintillation counting
2. Liquid scintillation counting

4.1   Solid Scintillation Counting

Instrumentation: *In solid scintillation counting, the fluorescing substance is held in a light-tight aluminum chamber. Radiations can penetrate through the aluminum and can cause excitation of solid crystals to emit photons of light. This fluorescence can be detected and quantified using a photomultiplier tube. In the photomultiplier tube, the electric pulse results from the conversion of light energy to electrical energy. This is directly proportional to the radioactive event as shown in Fig. 2.*

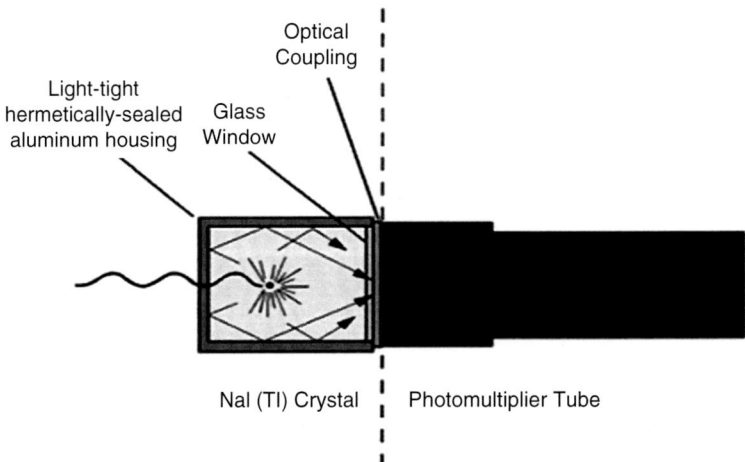

Optical Coupling

Light-tight hermetically-sealed aluminum housing   Glass Window

NaI (TI) Crystal   Photomultiplier Tube

Fig. 2 Solid scintillation counter

Crystal used for alpha isotopes: Zinc sulfide

Crystal used for beta isotopes: Anthracene

Crystal used for gamma isotopes: Sodium iodide

Advantages:

- Gamma isotopes have weak penetrating power; hence they rarely collide with molecules to cause excitation. In solid scintillation the atoms are densely packed in h crystal so the probability of collision is higher.

Disadvantages:

- Cannot detect weak beta emitters like 3H, 14C, and 35S as they are not able to pass through the wall of the counter.

4.2 Liquid Scintillation Counting

In liquid scintillation counting, the sample is mixed with a scintillation cocktail containing solvent and one or more fluors.

Instrumentation: In liquid scintillation counting, the radioactivity causes fluorescence of solvent. This emitted light in turn causes fluorescence of another compound called primary fluor. This emitted light in turn causes fluorescence of another compound called secondary fluor. This wavelength can be detected efficiently by the photomultiplier tube.

Advantages:

- Can detect weak beta emitters like 3H, 14C, and 35S
- More than one isotope can be detected at a time.
- Counting efficiency is high.

Disadvantages:

- It is costly.
- Quenching is the interference given by several substances in detecting actual fluorescence.

5   Ability to Expose Photographic Films (Autoradiography)

Principle: *Ionizing radiation acts upon photographic emulsions to produce a latent image.*

Instrumentation: *A radiation source and photographic emulsion consisting of silver halides embedded in solid phase such as gelatin are needed. Radiation converts silver halide to metallic silver which forms a latent image. This can be developed as a blackening of the film and can be stored as a permanent image.*

Uses: *Distribution of radioactivity in biological specimens*

6   Tracer Technique

It is technique in which one or more atoms of a chemical compound have been replaced by a radioisotope to trace the metabolic pathways. A radioactive tracer can also be used to track the distribution of a substance within a natural system such as a cell or tissue. Radioactive tracers form the basis of a variety of imaging systems, such as PET scans, SPECT scans, and technetium scans.

Principle:

It is based on the principle that a stable isotope is replaced by a radioisotope. The radioisotope is capable of emitting radiations that can be detected and analyzed. It is powerful than chemical reactions as they can be detected even in lower concentration as seen inside a cell. In general, isotopes of hydrogen, carbon, phosphorus, sulfur, and iodine have been used extensively to trace the path of biochemical reactions.

Uses:

1. *Tracing metabolic pathways*

    When a labeled chemical compound undergoes reaction inside the cell, one or more of the products will contain the radioactive label. Analysis of it provides detailed information on the mechanism of the chemical reaction.

2. *Distribution of the compound*

    Radioactive compound provides a means to construct an image showing the way in which it gets distributed in the organism.

Application:

1. In metabolism research, tritium (3H) and 14C-labeled glucose are commonly used to measure rates of glucose uptake, fatty acid synthesis, and other metabolic processes.

2. In medicine, tracers are applied in a number of tests, such as 99mTc in autoradiography and nuclear medicine, including single-photon emission computed tomography (SPECT), positron emission tomography (PET), and scintigraphy.

3. The urea breath test for *Helicobacter pylori* commonly used a dose of 14C-labeled urea to detect *H. pylori* infection.

7   Clinical and Biomedical Uses of Radioactive Substances

The radioactive substances can be used to study various metabolic pathways.

They are also used to diagnose and treat various diseases (e.g., cancer). Medical uses of few radioactive substances are given in the table below.

Table 1
Radioactive isotopes and their applications in biomedical research

Radioactive isotope	Uses
Carbon-14	Metabolism
Chromium-51	RBC studies
Cobalt-60	Sterilization of surgical instruments
Iodine-123,125	Diagnose thyroid disorders
Phosphorus-32,33	Molecular biology and genetic research
Sulfur-35	Molecular biology and genetic research
Tritium	Drug metabolism
Calcium- 47	Studying cellular function and bone formation
Cesium-137	Cancer treatment
Cobalt-57	Diagnosis of pernicious anemia
Copper-67	Treat cancer
Iodine-131	Treat thyroid disorders
Selenium-75	Protein studies
Technetium-99m	Organ imaging and blood flow studies
Xenon-133	Lung ventilation and blood flow

Clinical and Biomedical Uses of Radioactive Substances (Table 1):

8   Health Effects of Radiations

We already know that radioactive radiations are able to ionize and excite the molecules they encounter. The effects of radiation depend upon the duration and its type.

8.1   Effects

The effects of radiation are:

- *Generalized effects:* Biological effects are due to the ionization process that causes cell mutation. A given total dose will cause more damage if received in a shorter time period. A *fatal dose is 600 R.*

- *Acute somatic effects:* Relatively immediate effects to a person acutely exposed. Severity depends on dose. Death usually results from damage to bone marrow or intestinal wall. Acute *radio-dermatitis* is common in radiotherapy; chronic cases occur mostly in industry.

- *Critical organs*: Organs generally most susceptible to radiation damage include lymphocytes, bone marrow, gastrointestinal cells, gonads, and other fast-growing cells. The central nervous system is relatively resistant. Many nuclides concentrate in certain organs rather than being uniformly distributed over the body, and the organs may be particularly sensitive to radiation damage; e.g., isotopes of iodine concentrate in the thyroid gland. These organs are considered "critical" for the specific nuclide.

8.2 Basic Radiation Exposure Control Methods

The three basic ways to decrease the exposure to radiation are:

- Decrease time
- Increase distance
- Increase shielding

Time: Minimize time of exposure to minimize total dose. Rotate employees to restrict individual dose.

Distance: Maximize distance to source to maximize attenuation in air. The effect of distance can be estimated from equations.

Shielding: Minimize exposure by placing absorbing shield between worker and source.

Reference

1. Rajan K (2011) Analytical techniques in biochemistry and molecular biology. Springer, New York. eBook ISBN 978-1-4419-9785-2

Chapter 9

Microscopy and Specimen Preparation

Abstract

Biochemical analysis is frequently accompanied by microscopic examination of tissue, cell, or organelle preparations. Such examinations are used in many different applications, for example, to evaluate the integrity of samples during an experiment, to map the fine details of the spatial distribution of macromolecules within cells, and to directly measure biochemical events within living tissues. This chapter deals with light microscopy, scanning electron microscope, and tunneling electron microscope.

Key words Light microscopy, Scanning electron microscope, Tunneling electron microscope

Abbreviations

HVEM   High-voltage electron microscopy
SEM   Scanning electron microscope
TEM   Transmission electron microscope

Antonie van Leeuwenhoek (1632–1723) was the first person to observe and describe microorganisms accurately. In *1665* English physicist Robert Hooke looked at a sliver of cork through a microscope lens and noticed some "pores" or "cells" in it.

This is an optical instrument containing one or more lenses that produce an *enlarged image* of an object placed in the focal plane of the lens. Resolution limit: submicron particles approach the wavelength of visible light (400–700 nm).

- Eukaryotic cell – 20 μm
- Prokaryotic cell – 1–2 μm
- Nucleus of cell – 3–5 μm
- Mitochondria/chloroplast – 1–2 μm
- Ribosome – 20–30 nm
- Protein – 2–100 nm

Mahin Basha, *Analytical Techniques in Biochemistry*, Springer Protocols Handbooks,
https://doi.org/10.1007/978-1-0716-0134-1_9, © Springer Science+Business Media, LLC, part of Springer Nature 2020

Transmission: Beam of light *passes* through the sample. Samples are usually fine powder or thin slices (transparent), e.g., polarizing or petrographic microscope.

Reflection: Beam of light is *reflected* off the surface of sample materials, especially opaque ones, e.g., metallurgical or reflected light microscope.

Microscope Types:

- Bright field
- Stereo
- Phase contrast
- Differential interference contrast
- Fluorescence
- Confocal
- Electron
- Transmission
- Scanning
- Atomic force

Lenses and the Bending of Light: Light is refracted (bent) when passing from one medium to another.

Refractive index – The measure of how greatly a substance can slow down the velocity of light. Direction and magnitude of bending is determined by the refractive indices of the two media forming the interface.

Lenses focus light rays at a specific place called the focal point. Distance between the center of the lens and the focal point is the focal length.

- The strength of the lens is related to the focal length: short focal length ⇒ more magnification.

1   The Light Microscope

Basic Components of the Light Microscope: The simplest form of light microscope consists of a single glass lens mounted in a metal frame – a magnifying glass. Here the specimen requires very little preparation and is usually held close to the eye in the hand. Focusing of the region of interest is achieved by moving the lens and the specimen relative to one another. The source of light is usually the Sun or ambient indoor light. The detector is the human eye. The recording device is a hand drawing or an anecdote.

Compound Microscopes: All modern light microscopes are made up of more than one glass lens in combination. The major components are the condenser lens, the objective lens, and the eyepiece lens, and such instruments are therefore called compound microscopes.

The main components of the compound light microscope include a light source that is focused at the specimen by a condenser lens. Light that either passes through the specimen (transmitted light) or is reflected back from the specimen (reflected light) is focused by the objective lens into the eyepiece lens. The image is either viewed directly by eye in the eyepiece or it is most often projected onto a detector, for example, photographic film or, more likely, a digital camera. The images are displayed on the screen of a computer imaging system, stored in a digital format and reproduced using digital methods.

The part of the microscope that holds all of the components firmly in position is called the stand. There are two basic types of compound light microscope stand – an upright or an inverted microscope. The light source is below the condenser lens in the upright microscope and the objectives are above the specimen stage. This is the most commonly used format for viewing specimens. The inverted microscope is engineered so that the light source and the condenser lens are above the specimen stage and the objective lenses are beneath it. Moreover, the condenser and light source can often be swung out of the light path. This allows additional room for manipulating the specimen directly on the stage, for example, for the microinjection of macromolecules into tissue culture cells, for in vitro fertilization of eggs, or for viewing developing embryos over time.

The correct illumination of the specimen is critical for achieving high-quality images and photomicrographs. This is achieved using a light source. Typically light sources are mercury lamps, xenon lamps, lasers, or light-emitting diodes (LEDs).

Light from the light source passes into the condenser lens, which is mounted beneath the microscope stage in an upright microscope (and above the stage in an inverted microscope) in a bracket that can be raised and lowered for focusing. The condenser focuses light from the light source and illuminates the specimen with parallel beams of light. A correctly positioned condenser lens produces illumination that is uniformly bright and free from glare across the viewing area of the specimen (Koehler illumination). Condenser misalignment and an improperly adjusted condenser aperture diaphragm are major sources of poor images in the light microscope [1, 2].

The specimen stage is a mechanical device that is finely engineered to hold the specimen firmly in place. Any movement or vibration will be detrimental to the final image. The objective lens is responsible for producing the magnified image and can be the most expensive component of the light microscope in many different varieties, and there is a wealth of information inscribed on each one. This may include the manufacturer, magnification (4, 10, 20, 40, 60, 100), immersion requirements (air, oil, or water), coverslip thickness (usually 0.17 mm), and often more specialized optical properties of the lens. The numerical aperture (NA) is always marked on the lens. This is a number usually between 0.04 and 1.4. The NA is a measure of the ability of a lens to collect light from the specimen. Lenses with a low NA collect less light than those with a high NA [1, 2].

There are many types of microscopy:

– Bright-field microscope
– Dark-field microscope
– Phase-contrast microscope
– Fluorescence microscope
– Compound microscope
– Image formed by action of ≥ 2 lenses

1.1   The Bright-Field Microscope

In bright-field microscope, it produces a dark image against a brighter background and it has several objective lenses:

– Parfocal microscopes remain in focus when objectives are changed.

Total magnification is the product of the magnifications of the ocular lens and the objective lens.

Microscope Resolution: Lens has the ability to separate or distinguish small objects that are close together. Wavelength of light used is a major factor in resolution; for example, the shorter the wavelength, the greater the resolution.

1.2   The Dark-Field Microscope

The dark-field microscope produces a bright image of the object against a dark background; it is used to observe living, unstained preparations.

1.3   The Phase-Contrast Microscope

The phase-contrast microscope enhances the contrast between intracellular structures having slight differences in refractive index. Phase-contrast microscope is the excellent way to observe living cells. A phase ring in condenser allows a cylinder of light through while in phase. Light that is unaltered hits the phase ring in the lens and is excluded. Light that is slightly altered by passing through different refractive indices is allowed through.

Light passing through cellular structures such as chromosomes or mitochondria is retarded because they have a higher refractive index than the surrounding medium. Elements of lower refractive index advance the wave. Much of the background light is removed and light that constructively or destructively interfered is let through with enhanced contrast. It visualizes differences in refractive index of different parts of a specimen relative to the unaltered light.

1.4 The Differential Interference Contrast Microscope	The differential interference contrast microscope creates image by detecting differences in refractive indices and thickness of different parts of the specimen. It is an excellent way to observe living cells.
1.5 The Fluorescence Microscope	Fluorescent dye – Molecules that absorb light of one wavelength and then re-emit it at a longer wavelength. The fluorescence microscope exposes specimen to ultraviolet, violet, or blue light. Specimens are usually stained with fluorochromes; it shows a bright image of the object resulting from the fluorescent light emitted by the specimen.

2   Preparation and Staining of Specimens

Staining increases the visibility of specimens, accentuates specific morphological features, and preserves specimens.

2.1 Fixation	It is the process by which internal and external structures are preserved and fixed in position. It is a process wherein an organism is killed and firmly attached to a microscope slide.

Heat fixing: It preserves overall morphology but not internal structures.

Chemical fixing: It protects fine cellular substructure and morphology of larger, more delicate organisms. |
| **2.2 Dyes** | Dyes make internal and external structures of cells more visible by increasing contrast with the background. It has two common features:

– Chromophore groups: Chemical groups with conjugated double bonds.

– Dye gives its color to the specimen.
   • *Ability to bind cells* |
| **2.3 Simple Staining** | In this staining a single staining agent is used and basic dyes are frequently used mostly dyes with positive charges, e.g., crystal violet. |

2.4 Differential Staining

It divides microorganisms into groups based on their staining properties, e.g., Gram stain and acid-fast stain.

2.4.1 Gram Staining

It is the most widely used differential staining procedure. It divides bacteria into two groups based on differences in cell wall structure as shown in Fig. 1.

2.4.2 Acid-Fast Staining

This staining is particularly useful for staining members of the genus *Mycobacterium*, e.g., *Mycobacterium tuberculosis* that causes tuberculosis and *Mycobacterium leprae* that causes leprosy,

- High lipid content in cell walls is responsible for their staining characteristics.

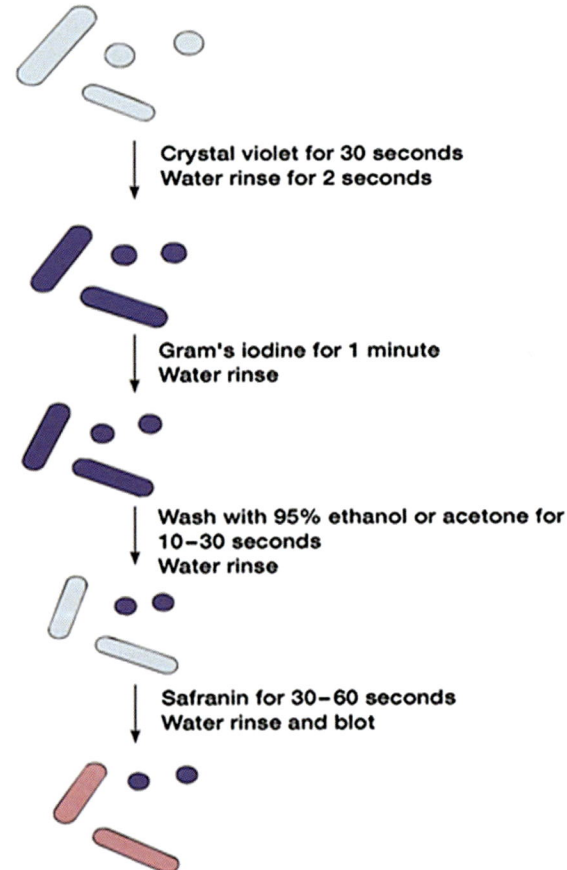

Fig. 1 Gram staining

2.4.3 Staining Specific Structures

Negative staining: Often used to visualize capsules surrounding bacteria. Capsules are colorless against a stained background.
Spore staining: Double staining technique.

– Bacterial endospore is one color and vegetative cell is a different color.

Flagella staining: Mordant is applied to increase thickness of flagella.

3 Electron Microscopy

Beams of electrons are used to produce images. Wavelength of electron beam is much shorter than light, resulting in much higher resolution as shown in Table 1.

Scanning electron microscopy is used for inspecting topographies of specimens at very high magnifications using a piece of equipment called the scanning electron microscope. SEM magnifications can go to more than 300,000 X, but most semiconductor manufacturing applications require magnifications of less than 3,000 X only. SEM inspection is often used in the analysis of die/package cracks and fracture surfaces, bond failures, and physical defects on the die or package surface.

4 Principles

Electron microscopy is used when the greatest resolution is required and when the living state can be ignored. The images produced in an electron microscope reveal the ultra-structure of cells. There are two different types of electron microscope – the

Table 1
Probes used in electron microscopy

Probes used	
• *Visible light*	• *Electron*
– Optical microscopy (OM)	– Scanning electron microscopy (SEM)
• *X-ray*	– Transmission electron microscopy (SEM)
– X-ray diffraction (XD)	– Electron holography (EH)
– X-ray photo electron spectroscopy (XPS)	– Electron diffraction (ED)
• *Neutron*	– Electron energy loss spectroscopy (EELS)
– Neutron diffraction (ND)	– Energy dispersive x-ray spectroscopy (EDS)
• *Ion*	– Auger electron spectroscopy (AES)
– Secondary ion mass spectroscopy (SIMS)	
– Cleaning and thinning samples	

transmission electron microscope (TEM) and the scanning electron microscope (SEM). In the TEM, electrons that pass through the specimen are imaged. In the SEM, electrons that are reflected back from the specimen (secondary electrons) are collected, and the surfaces of specimens are imaged.

The equivalent of the light source in an electron microscope is the electron gun. When a high voltage of between 40,000 and 100,000 volts (the accelerating voltage) is passed between the cathode and the anode, a tungsten filament emits electrons. The negatively charged electrons pass through a hole in the anode forming an electron beam. The beam of electrons passes through a stack of electromagnetic lenses (the column). Focusing of the electron beam is achieved by changing the voltage across the electromagnetic lenses. When the electron beam passes through the specimen, some of the electrons are scattered while others are focused by the projector lens onto a phosphorescent screen or recorded using photographic film or a digital camera. The electrons have limited penetration power which means that specimens must be thin (50–100 nm) to allow them to pass through.

Thicker specimens can be viewed by using a higher accelerating voltage; for example, in the high-voltage electron microscope (HVEM), an accelerating voltage of 1,000,000 V is used, while in the intermediate-voltage electron microscope (IVEM), an accelerating voltage of around 400,000 V is used. Here stereo images are made by collecting two images at 8–10 tilt angles. Such images are useful in assessing the 3D relationships of organelles within cells when viewed in a stereoscope or with a digital stereo projection system [1, 2].

5   Preparation of Specimens

Contrast in the EM depends on atomic number; the higher the atomic number, the greater the scattering and the contrast. Thus heavy metals are used to add contrast in the EM, for example, uranium, lead, and osmium. Labeled structures appear black or electron dense in the image. All of the water has to be removed from any biological specimen before it can be imaged in the EM. This is because the electron beam can only be produced and focused in a vacuum. The major drawback of EM observation of biological specimens therefore is the non-physiological conditions necessary for their observation. Nevertheless, the improved resolution afforded by the EM has provided much information about biological structures and biochemical events within cells that could not have been collected using any other microscopic technique [1, 2].

Extensive specimen preparation is required for EM analysis, and for this reason there can be issues of interpreting the images because of artifacts from specimen preparation. For example, specimens have been traditionally prepared for the TEM by fixation in glutaraldehyde to cross-link proteins followed by osmium tetroxide to fix and stain lipid membranes. This is followed by dehydration in a series of alcohols to remove the water and then embedding in a plastic such as Epon for thin sectioning. For the SEM, samples are fixed in glutaraldehyde, dehydrated through a series of solvents, and dried completely either in air or by critical point drying. This method removes all of the water from the specimen instantly and avoids surface tension in the drying process thereby avoiding artifacts of drying. The specimens are then mounted onto a special metal holder or stub and coated with a thin layer of gold before viewing in the SEM (Fig. 4.22; see also color section). Surfaces can also be viewed in the TEM using either negative stains or carbon replicas of air-dried specimens.

6 Characteristics of SEM

Topography: The surface features of an object or how it looks or its texture; there is a direct relationship between these features and materials properties.

Morphology: The shape and size of the particles making up the object; there is a direct relationship between these structures and materials properties.

Composition: The elements and compounds that the object is composed of and the relative amounts of them; there is a direct relationship between composition and materials properties.

Crystallography: The arrangement of atoms in the object; there is a direct relationship between these arrangements and materials properties.

1. The virtual source at the top represents the electron gun. It produces a stream of monochromatic electrons.

2. The stream is condensed by the first condenser lens (usually controlled by the coarse probe current knob). This lens is used to both form the beam and limit the amount of current in the beam. It works in conjunction with the condenser aperture to eliminate the high-angle electrons from the beam.

3. The beam is then constricted by the condenser aperture, eliminating some high-angle electrons.

4. The second condenser lens forms the electrons into a thin, tight, coherent beam and is usually controlled by the fine probe current knob.

5. A user selectable objective aperture further eliminates high-angle electrons from the beam.

6. A set of coils then scan or sweep the beam in a grid fashion, dwelling on points for a period of time determined by the scan speed (usually in the microsecond range).

7. The final lens, the objective, focuses the scanning beam onto the part of the specimen desired.

8. When the beam strikes the sample, interactions occur inside the sample and are detected with various instruments.

9. Before the beam moves to its next dwell point, these instruments count the number of electron interactions and display a pixel on a CRT whose intensity is determined by the number (the more reactions there are, the brighter the pixel).

10. The process is repeated until the grid scan is finished; the entire pattern can be scanned 30 times/sec.

During SEM inspection, a beam of electrons is focused on a spot volume of the specimen, resulting in the transfer of energy to the spot. These bombarding electrons, also referred to as primary electrons, dislodge electrons from the specimen itself. The dislodged electrons, also known as secondary electrons, are attracted and collected by a positively biased grid or detector and then translated into a signal as shown in Fig. 2.

To produce the SEM image, the electron beam is swept across the area being inspected, producing many such signals. These signals are then amplified, analyzed, and translated into images of the topography being inspected. Finally, the image is shown on a CRT.

The energy of the primary electrons determines the quantity of secondary electrons collected during inspection. The emission of

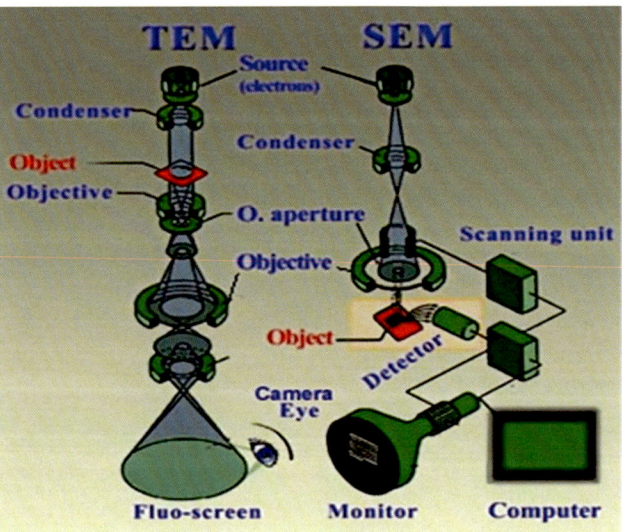

Fig. 2 SEM and TEM

secondary electrons from the specimen increases as the energy of the primary electron beam increases, until a certain limit is reached. Beyond this limit, the collected secondary electrons diminish as the energy of the primary beam is increased, because the primary beam is already activating electrons deep below the surface of the specimen. Electrons coming from such depths usually recombine before reaching the surface for emission.

Aside from secondary electrons, the primary electron beam results in the emission of backscattered (or reflected) electrons from the specimen. Backscattered electrons possess more energy than secondary electrons and have a definite direction. As such, they cannot be collected by a secondary electron detector, unless the detector is directly in their path of travel. All emissions above 50 eV are considered to be backscattered electrons.

Backscattered electron imaging is useful in distinguishing one material from another, since the yield of the collected backscattered electrons increases monotonically with the specimen's atomic number. Backscattered imaging can distinguish elements with atomic number differences of at least three, i.e., materials with atomic number differences of at least three would appear with good contrast on the image. For example, inspecting the remaining Au on an Al bond pad after its Au ball bond has lifted off would be easier using backscattered imaging, since the Au islets would stand out from the Al background.

A SEM may be equipped with an EDX analysis system to enable it to perform compositional analysis on specimens. EDX analysis is useful in identifying materials and contaminants, as well as estimating their relative concentrations on the surface of the specimen.

7 TEM

In a conventional transmission electron microscope, a thin specimen is irradiated with an electron beam of uniform current density. Electrons are emitted from the electron gun and illuminate the specimen through a two- or three-stage condenser lens system. Objective lens provides the formation of either image or diffraction pattern of the specimen. The electron intensity distribution behind the specimen is magnified with a three- or four-stage lens system and viewed on a fluorescent screen. The image can be recorded by direct exposure of a photographic emulsion or an image plate or digitally by a CCD camera as shown in Fig. 2.

The acceleration voltage of up-to-date routine instruments is 120–200 kV. Medium-voltage instruments work at 200–500 kV to provide a better transmission and resolution, and in high-voltage electron microscopy (HVEM), the acceleration voltage is in the range of 500 kV to 3 MV. Acceleration voltage determines the velocity, the wavelength, and hence the resolution (ability to distinguish the neighboring microstructural features) of the microscope.

Depending on the aim of the investigation and configuration of the microscope, transmission electron microscopy can be categorized as:

- Conventional transmission electron microscopy
- High-resolution electron microscopy
- Analytical electron microscopy
- Energy-filtering electron microscopy
- High-voltage electron microscopy
- Dedicated scanning transmission electron microscopy

TEM consists of the following major parts:

1. The illumination system
2. The image-forming system
3. The protective system
4. Apertures

8 Applications

- Image morphology of samples, e.g., view sections of material, fine powders suspended on a thin film, small whole organisms such as viruses or bacteria, and frozen solutions.
- Tilt a sample and collect a series of images to construct a three-dimensional image.
- Analyze the composition and some bonding differences (through contrast and by using spectroscopy techniques such as microanalysis and electron energy loss).
- Physically manipulate samples while viewing them, such as indenting or compressing them to measure mechanical properties (only when holders specialized for these techniques are available).
- View frozen material (in a TEM with a cryostage).
- Generate characteristic X-rays from samples for microanalysis.
- Acquire electron diffraction patterns (using the physics of Bragg diffraction).
- Perform electron energy loss spectroscopy of the beam passing through a sample to determine sample composition or the bonding states of atoms in the sample.

References

1. Rajan K (2011) Analytical techniques in biochemistry and molecular biology. Springer, New York. eBook ISBN978-1-4419-9785-2

2. Wilson K, Walker J (2010) Principles and techniques of biochemistry and molecular biology. Cambridge University Press, Cambridge. ISBN 978-0-521-73167-6

INDEX

Mahin Basha, *Analytical Techniques in Biochemistry*, Springer Protocols Handbooks,
https://doi.org/10.1007/978-1-0716-0134-1, © Springer Science+Business Media, LLC, part of Springer Nature 2020

Zeitfracht Medien GmbH
Ferdinand-Jühlke-Straße 7
99095 Erfurt, Deutschland
produktsicherheit@kolibri360.de